Lightning Induced Overvoltage and Current on Distribution Power Systems

Behrooz Vahidi · Amin Foroughi Nematolahi

Lightning Induced Overvoltage and Current on Distribution Power Systems

Springer

Behrooz Vahidi
Department of Electrical Engineering
Amirkabir University of Technology
Tehran, Iran

Amin Foroughi Nematolahi
Department of Electrical Engineering
Amirkabir University of Technology
Tehran, Iran

ISBN 978-981-95-3711-2 ISBN 978-981-95-3712-9 (eBook)
https://doi.org/10.1007/978-981-95-3712-9

This Springer imprint is published by the registered company Springer Nature Singapore Pte Ltd.
The registered company address is: 152 Beach Road, #21-01/04 Gateway East, Singapore 189721, Singapore

Preface

In recent decades electronics devices are widely used and due to influence of direct and indirect lightning on normal function of these devices, protection of them against lightning is important. Lightning is a phenomenon which has effect on power quality and system reliability of the power distribution systems. Due to height and location of the power distribution lines indirect lighting stroke has more influence on distribution system with respect to direct stroke. In different geographical area Lightning will be happened on cloudy days and in the specific months of year. Direct and indirect strokes can put power system in a critical situation in distribution system the induced overvoltage can be more than BIL of system therefore damage due to lightning will increase.

Therefore, to whom is this book directed? It will truly be of use to postgraduate students and power system engineers.

This book is divided into seven chapters. Chapter 1 shortly deals with physics of lightning and lightning protection of power system. Chapter 2 dicusses the effect of lightning on the power systems and different protection divices. Chapter 3 is focused on literature review on the lightning induceded voltage due to indirect lightning stroke. Chapter 4 is about identifying lightning channel-base current characteristics and finding Heidler function parameters which will be used in induced voltage and current computation. Chapter 5 is about analyzing the effect of inclined return stroke channel on the induced voltage on the overhead distribution line which can be compared with results of traditional vertical return stroke channel. Chapter 6 is analyzing the performance of overhead distribution line against indirect lightning by considering ground with two vertical or horizontal layers. Finally Chap. 7 is about analyzing induced current in the shield of underground cable due to inclined return stroke channel by using 3D FEM.

The goal of this book is to educate the postgraduate students and engineers about principals and modeling of indirect lightning stroke.

The authors' special thanks go to all readers in advance who will give us feedback on the book.

Tehran, Iran
June 2025

Behrooz Vahidi
Amin Foroughi Nematolahi

Acknowledgments The authors wish to acknowledge the assistance provided by Dr. A. Teymouri for organizing the chapter's format.

Contents

Chapter 1
Lightning Physics and the Lightning Protection

Abstract In recent decades electronics devices are widely used and due to influence of direct and indirect lightning on normal function of these devices, protection of them against lightning is important. Lightning is a phenomenon which has effect on power quality and system reliability of the power distribution systems. Due to height and location of the power distribution lines indirect lighting stroke has more influence on distribution system with respect to direct stroke. Chapter one shortly deals with physics of lightning and lightning protection of power system. In this chapter at frist we discuss about physics of lightning and different types of linghtning, then computation of lightning stroke to overhead transmission lines are dicussed.

1.1 Introduction

In recent decades, electronic devices have been widely used, and due to the influence of direct and indirect lightning on the normal function of these devices, protecting them against lightning is essential. Lightning is a phenomenon that affects the power quality and reliability of power distribution systems. Due to the height and location of the power distribution lines, the indirect lighting stroke has more influence on the distribution system than the direct stroke.

In different geographical areas, Lightning will happen on cloudy days and in specific months of the year. Direct and indirect strokes can put the power system in a critical situation in the distribution system. The induced overvoltage can be more than the BIL of the system; therefore, damage due to lightning will increase.

1.2 Lightning Physics

Lightning can be considered as a high-current transient electric discharge with a path of several kilometers when electric charges of some parts of the atmosphere are enough to produce high electric fields to break down the air, resulting in lightning.

B. Vahidi and A. F. Nematolahi, *Lightning Induced Overvoltage and Current on Distribution Power Systems*, https://doi.org/10.1007/978-981-95-3712-9_1

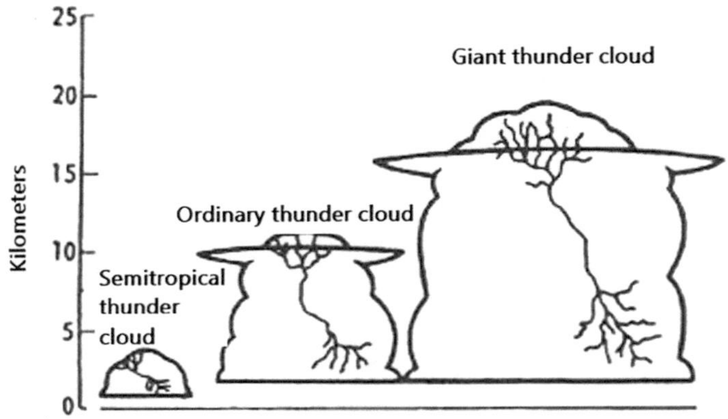

Fig. 1.1 A different kind of thunder cloud

One of the famous thunder clouds is cumulonimbus. Lightning can occur in snowy weather, in the clouds over the volcanoes during an eruption. Lightning can even happen in clear weather [1–5].

In the following conditions, a thunder cloud is shaped: in cloudy air, at high levels, thick air, and moist, warmer air at lower heights. The warmer air goes up in a heavy draft form to shape clouds, and the cold, dense air descends. Sometimes frigid polar air overruns the places with very hot ground, which transfers its heat to the atmosphere or warm air, and then the thunder cloud is shaped. Figure 1.1 shows a different-sized thunder cloud. The thundercloud height typically is from 8 to 12 km; however, for a specific geographic position, typical values can only be presented.

1.2.1 Cloud Electrification Theory

At present, there is no unique theory for the electrification of thunder clouds. At least there are two kinds of theories, one of them focuses on the dropping of the precipitation particles for charge separation in the clouds, and the other one focuses on the convection behavior of cloudy air in order to transfer the charges that are derived externally in a more or less organized manner [5].

A deep explanation for cloud electrification can be found in references [6–11].

1.2.2 Precipitation-Powered Processes

Many researchers believe that the selective separation and transportation of negative charge toward the ground in thunderclouds is a process that is done by precipitation

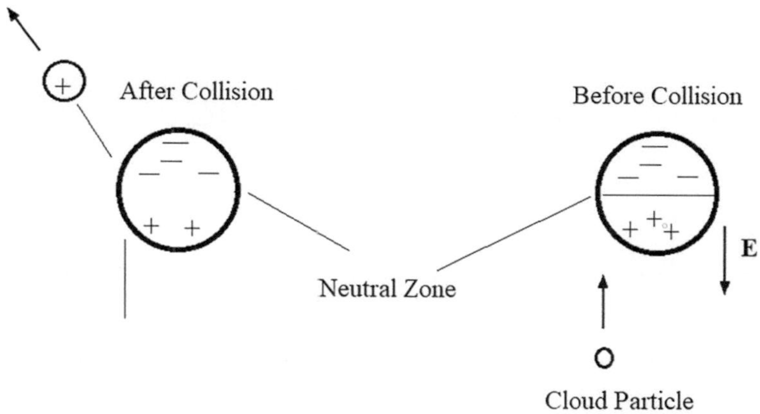

Fig. 1.2 The induction charge transfer process

particles that fall [9–11]. Figure 1.2 shows the process of induction charge transportation. This figure shows the method that charges are postulated as being separated in elastic collisions that occur among particles of cloud and a vertical electric field polarized falling precipitation.

1.2.3 Convective Electrification Mechanism

Grenet [12] suggested the proposed mechanism. After that, it was recommended by Vonnegut [13] that the motions in the convective method in the clouds transport and accumulate the particles of cloud that are electrified in opposition to the electric field force. The intensity of electric fields and their energy will increase due to the accumulated negative and positive charges. The negative and positive charges, which are needed for this mechanism, are derived from the clouds. One ampere of conducting current from the upper clear atmosphere over the cloud flows and will transfer a small number of ions with opposing polarity to the higher parts of the cloud, then these ions are quickly immobilized by attaching to particles of the cloud.

As a result, a thick negative polarity space charge layer builds up on the surface of the cloud, which is then caused by the convective circulation and accumulates as charge pockets inside the cloud. At the bottom of the cloud, by point-discharge, the ions with positive polarity are released at about 1 A rate and are moved by the upflow of air into the cloud, where these ions attach to the particles of the cloud and accumulate as regions with charge in the higher level part of the cloud. Figure 1.3 shows the convection mechanism of electrification.

This method of electrification has positive feedback and is intensifying by itself. If an electric field or space charge exists, it is initiated at the beginning phase of the cloud's production. This primary electrification happens due to different sources,

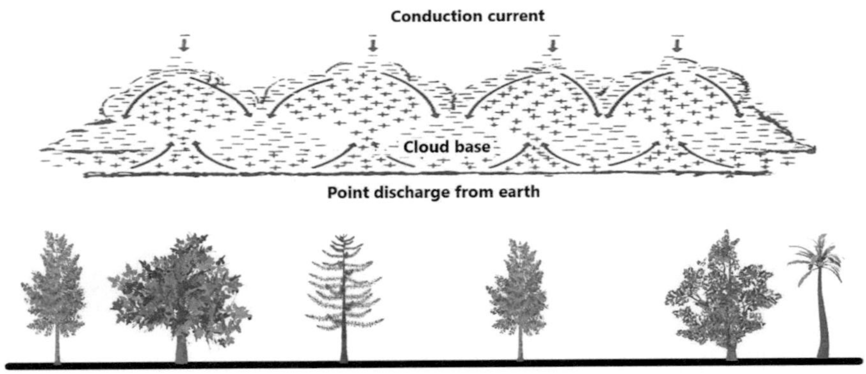

Fig. 1.3 The convection mechanism of electrification

such as fair-weather electricity, electrified ocean spray, dust flow, dropping charged particles, or an electrified cloud that is near this cloud.

A mechanism, perhaps one of the above-mentioned mechanisms or other mechanisms that are introduced by researchers in this field, could be the reason for the thunderstorm electrification.

1.2.4 Different Types of Lightning Flash

The negative charges are accumulated in the lower parts of thunderclouds, but the higher parts of thunderclouds contain positive charges [9–11]. The polarity of cloud discharge to earth defines the type of flash (negative or positive). In a positive flash, positive current flows toward the ground, and negative currents flow toward the ground in a negative flash. As a downward leader starts toward the earth from the thundercloud, an upward leader starts from the ground toward the cloud. The progress of an upward leader from the ground (or a grounded object) that reaches the downward leader and discharges it to ground.

At a point with an electric field inside the cloud, a leader can initiate. This process may happen either between centers of negative and positive charges inside a cloud or between centers of positive (or negative) charge centers in the cloud and its induced counter charge in the earth.

The first one can lead to a flash inside the cloud or can proceed toward earth and make a ground flash.

In the second one, the process starts with a leader stroke from the charged cloud to the charge, which is induced in the earth. The biggest field strength may then be produced at the bottom of the cloud or on a higher grounded object. This process can create a downward flash from cloud to ground or to an upward leader which starts from the grounded object toward the cloud [14].

Different lightning flash types are as follows:

Designation: Downward Negative Leader

A leader toward the ground that starts from a negatively charged cloud before discharge. This happens in the open area where there are no tall objects. This leader contains a negative charge and a negative current. If the leader cannot touch the ground (air discharge), there is no return stroke; this is a cloud discharge. Figure 1.4 shows this case.

Designation: Downward Negative Flash

When a negative leader that is toward the ground reaches the ground, a fast upward-moving return stroke is produced, and the return stroke discharges the leader and part of the cloud charge to ground. If this process occurs once, a single-stroke flash call will be produced, but if it repeats, a multi-stroke flash will be created.

Figure 1.5 shows the downward negative flash.

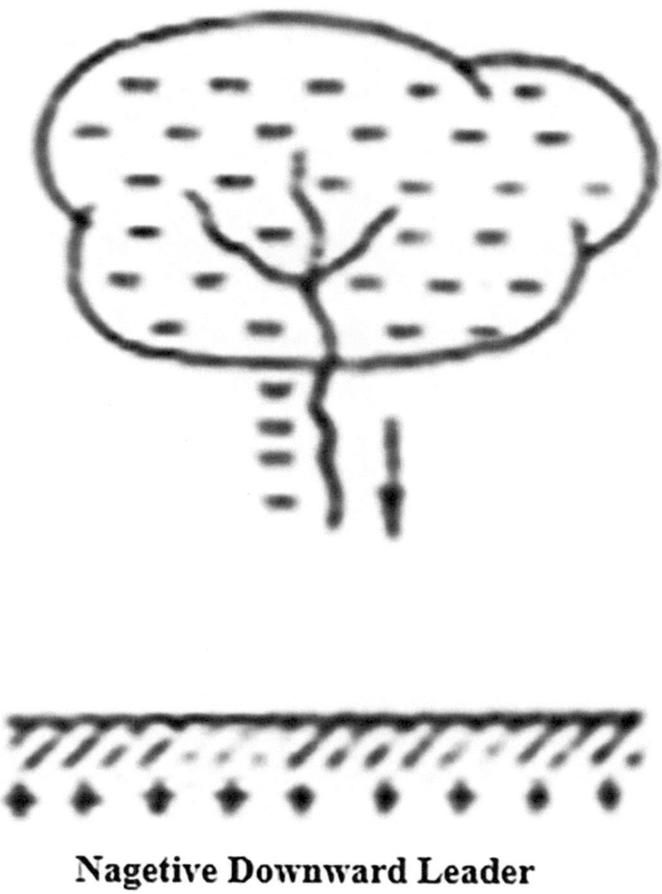

Fig. 1.4 The downward negative leader

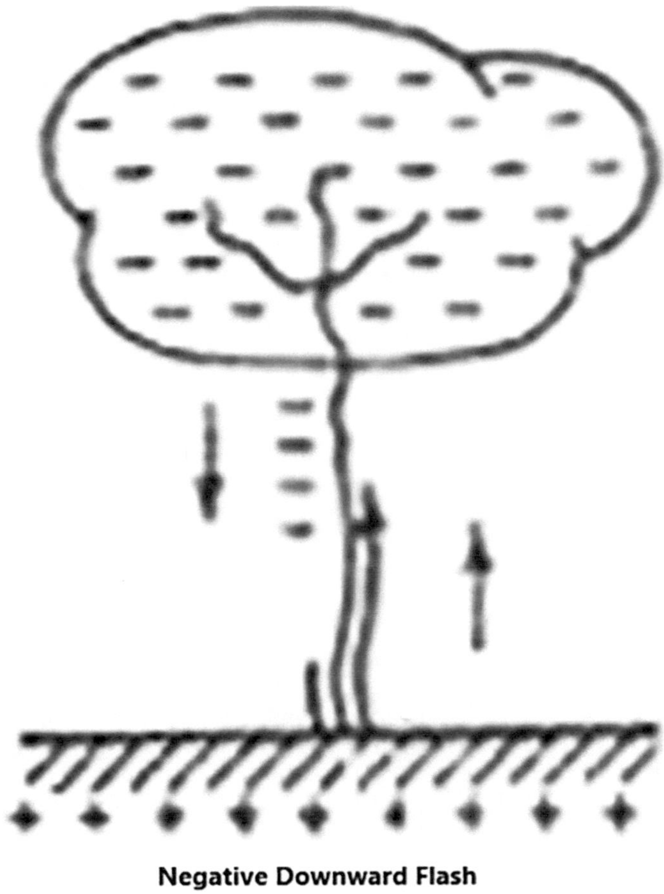

Fig. 1.5 The downward negative flash

Designation: Upward Positive Leader/Continuous Negative Flash

The discharge starts from a tall object on the earth by a leader toward a negative cloud from the top of a tower or the top of a mountain. The leader's charge is positive (the top of the object acts as the anode); it is a positive leader. Cloud charge flows to the earth, and therefore, the tower currents are negative. Figure 1.6 shows the positive upward leader/negative continuous flash.

Designation: Upward Positive Leader/Multiple Negative Flash

The initial stage of this type is as shown in Fig. 1.6, and it continues with other strokes, the downward leader, and upward return stroke of each are as shown in Fig. 1.5. Figure 1.7 shows an upward positive /multiple negative flash.

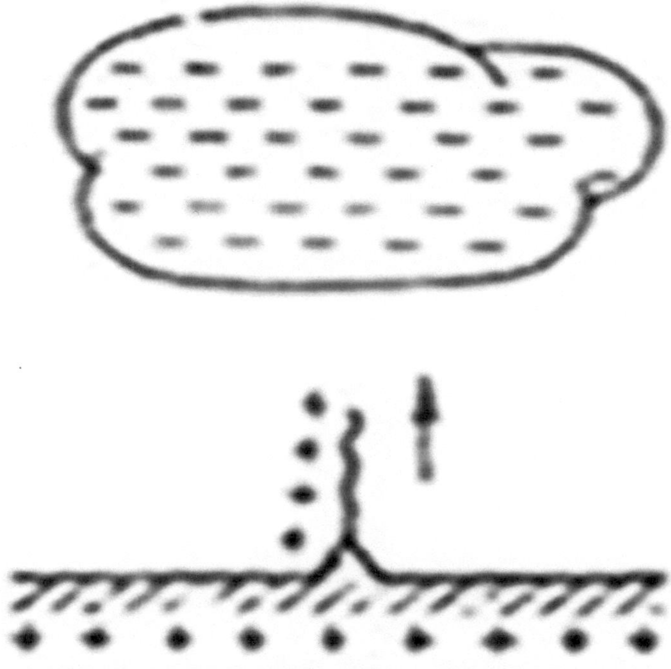

Fig. 1.6 The upward positive leader/continuous negative flash

Designation: Positive Downward Leader

This type corresponds to the type of Fig. 1.4, but under a cloud with a positive charge. The leader's charge and its current are positive. The leader cannot reach the ground, and flows of displacement current occur in the earth. This is an intra-cloud flash. Figure 1.8 shows the positive downward leader.

Designation: Downward Positive Flash

When the positive downward leader reaches the earth, it gives rise to a positive upward return stroke by which the leader and part of the cloud are discharged. This type is scarce in mountainous regions. Figure 1.9 shows the positive downward flash.

Designation: Upward Negative Leader/Positive Continuous Current Flash, or, Positive continuous-current Flash

By initiating a leader from a tall object on the earth toward the cloud beneath a cloud with a positive charge. The charge of the leader is negative (The tip of the object acts as the cathode). The following charge and current to ground are positive, which starts a long-duration current, described as a continuous current. Figure 1.10 shows

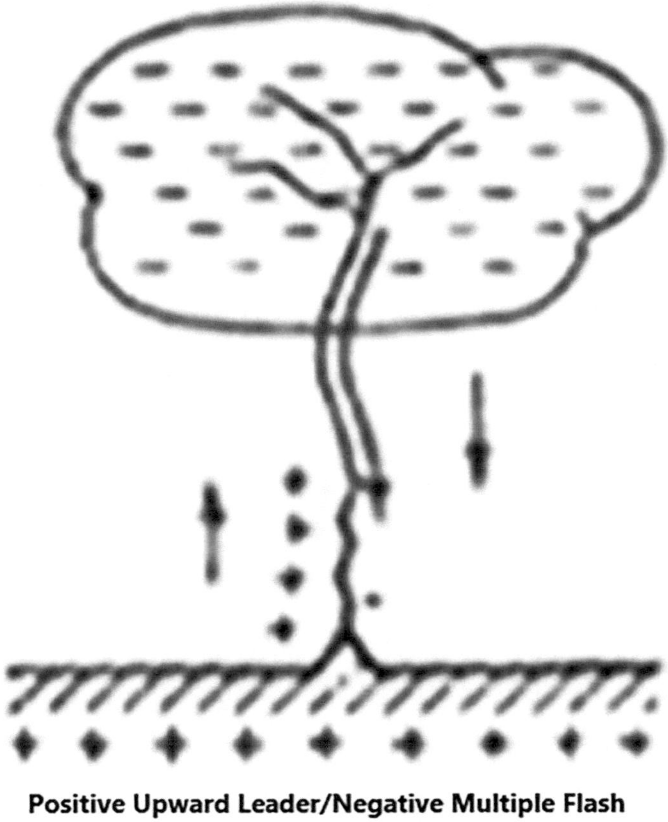

Fig. 1.7 The upward positive leader/multiple negative flash

the upward negative leader/positive continuous-current flash, or positive continuous-current flash.

Designation: Upward Negative Leader/Positive Impulse-current Flash

This type starts as type Fig. 1.10, but after four to twenty milliseconds, an impulsive, severe downward positive discharge follows the upward leader, which is regarded as a return stroke. Berger observed this type [15]. The progress of an upward negative leader in the form of a long "connecting leader", into an existing intercloud flash, which, first, causes the transient field by which the upward leader was initiated, and then it is discharged to ground through this "connecting leader". Figure 1.11 shows an upward negative leader/positive impulse-current flash.

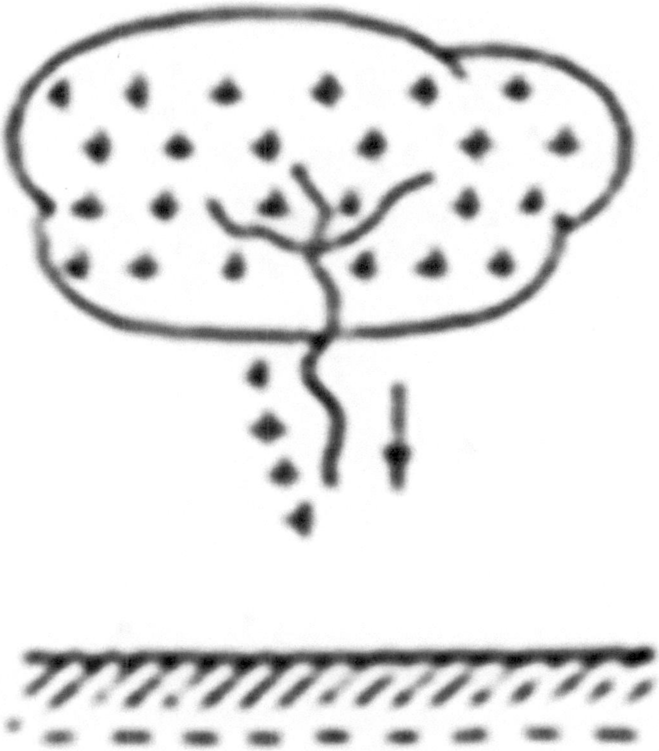

Fig. 1.8 The downward positive leader

1.2.5 Leader and Return Stroke

Stepped Leader

The initial breakdown yields a leader inside the cloud, which is bidirectional. The section of the channel that is positive moves toward the center of the negative charge, and the negative part of the channel moves toward the ground. Most of the time, the negative part of the leader's motion is in stepped shape, forming toward the earth, and it is called the stepped leader [11, 16].

As we mentioned, it is named the stepped leader due to its motion, which is toward the ground with 50 m length luminous steps, and there is about 50 microseconds pause time between steps. The stepped leader channel is not luminous during the pause time, or there is not enough lamination that can be recorded in photographic film with a boys' camera or streak-camera techniques. The stepped leader velocity (average value) during its trip to ground is typically 1.5×10^5 m/s. Figure 1.12 shows a stepped leader.

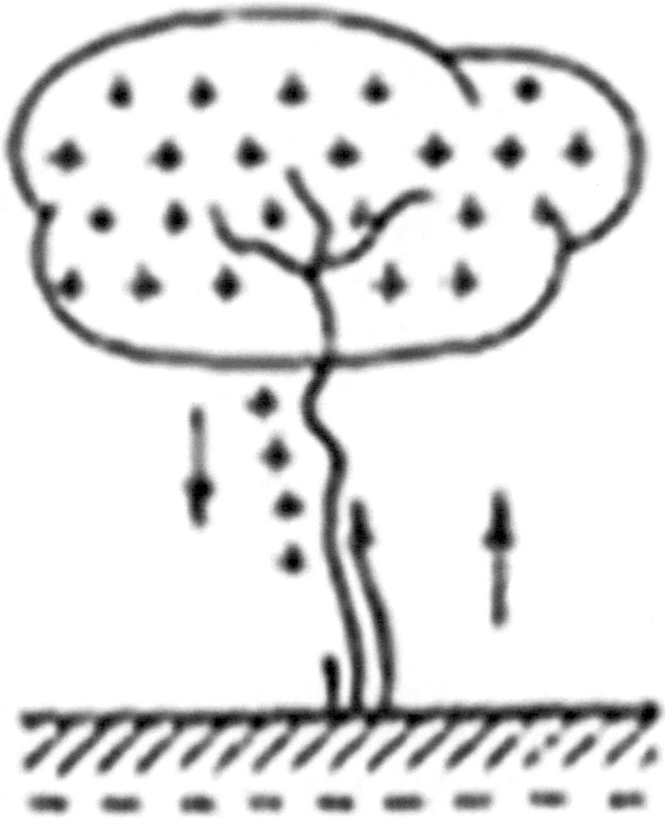

Fig. 1.9 The downward positive flash

Connecting Leader

When a leader with a step shape goes toward the earth, then the electric field at the earth's surface increases step by step. As the downward leader reaches several hundred meters to earth, at the tips of earthed objects, electric fields reach a high level, which discharges from the tips of these objects, and these discharges are called connecting leaders (Fig. 1.12). These discharges move toward the downward leaders. The connecting leader length is around a few tens of meters [16]. The connecting leaders' lengths depend on different parameters, such as:

(i) For objects with a higher length, the connecting leader is longer.
(ii) For an object with a given length, the connecting leader is longer for a stepped leader with higher charges.
(iii) The geometry of the object that the connecting leader starts.

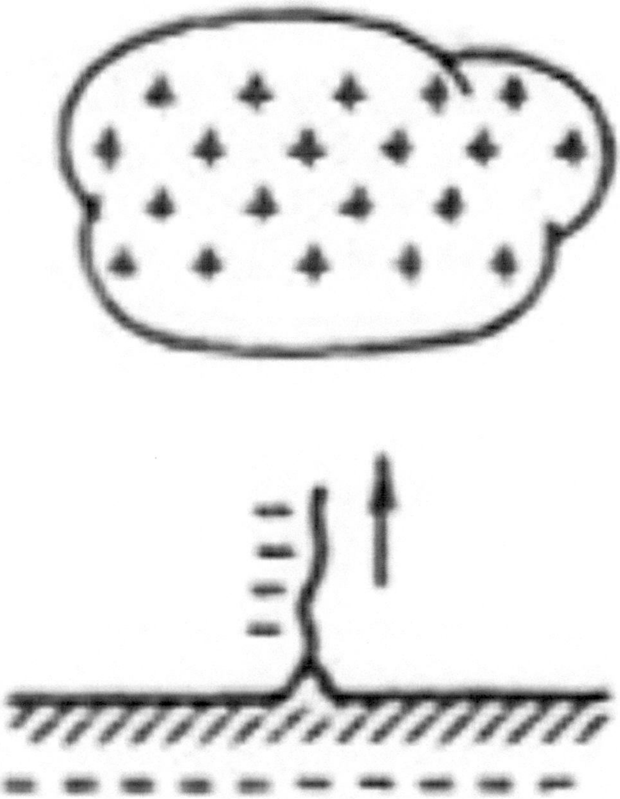

Fig. 1.10 The upward negative leader/positive continuous-current flash, or positive continuous-current flash

(iv) The lateral distance between the object and the stepped leader.
(v) The object should be able to prepare enough current for propagation.

Striking Distance

When the stepped leader goes toward the earth, an upward connecting leader is initiated and moves toward the downward leader as much as these two leaders reach closer the electric field difference between the tips of these leaders increases at a distance between tips of them the final jump occurs, and two leader touch toghter, this distance is called striking distance. Figure 1.13 is a schematic description of striking distance.

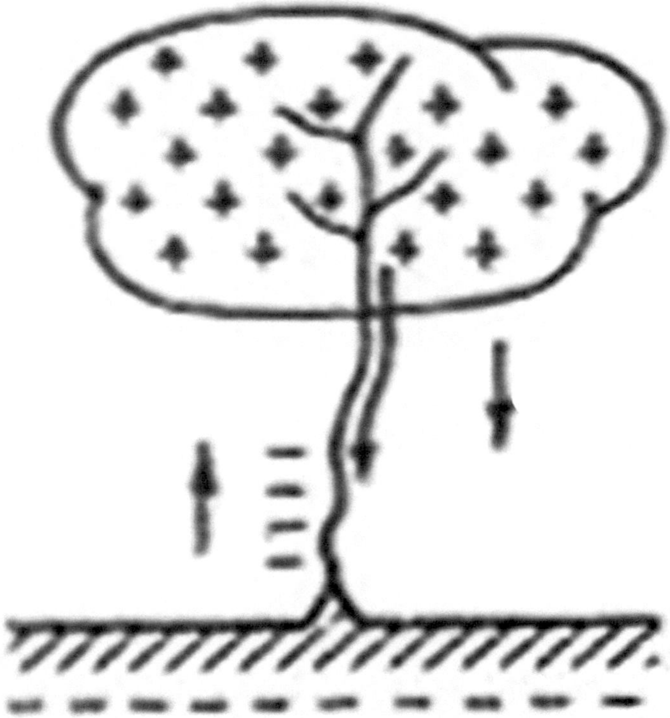

Fig. 1.11 The upward negative leader/positive impulse-current flash

Return Stroke

When the downward leader and upward leader from an earthed object touch each other, the lightning strikes the object [16]. After the connection between downward and upward leaders, a wave will travel from earth to cloud, which is associated with a luminous event that travels toward the cloud, and this event is called the return stroke.

A very luminous return stroke is supported by a leader channel that works the same as a transmission line (lossy and nonlinear). The velocity of the wave front of the return stroke is about one-tenth to one-third of light speed [1]. Usually, the first stroke in a flash is downward and with different branches, as is the preceding stepped leader. Subsequent return strokes following dart leaders show little branching. The propagation velocity (average) of first return stroke is slower, with longer times to peak current, and increasing rate of current at ground is slower (Fig. 1.12).

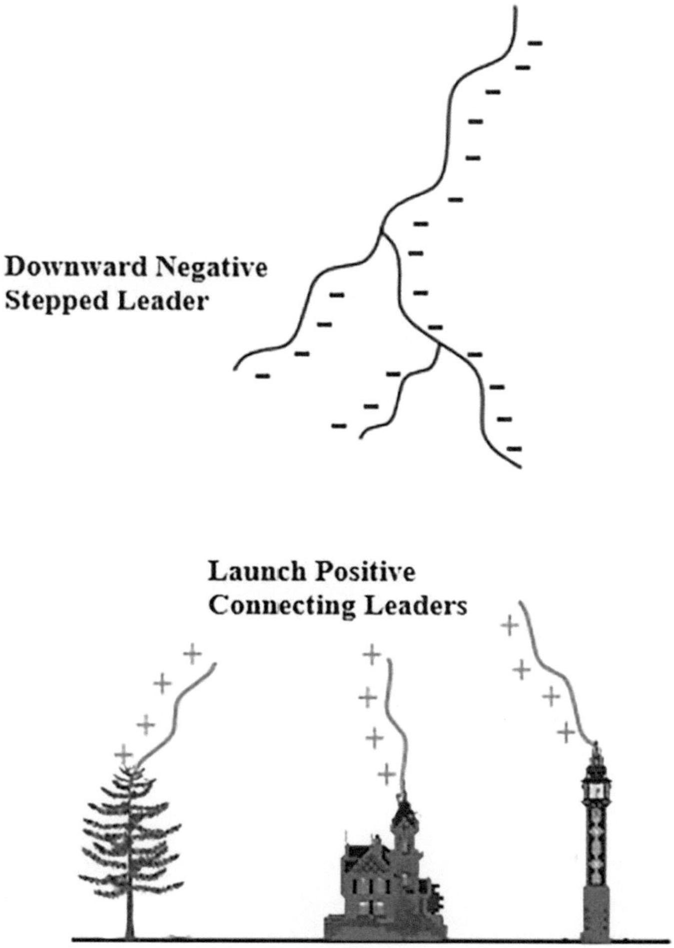

Fig. 1.12 Stepped leader

Dart Leader

By ending the flow of the lightning current, the lightning flash will probably end. Otherwise, if there are some more charges at the top of the channel, more strokes can happen in a flash, and this kind of flash is called multiple-stroke flash. After each stroke, usually each subsequent stroke appears to drain charge from the N region of the upper area of the cloud. The charge is provided between strokes by the action that calls the K steamer and J streamer behavior, electric discharge at the higher end of the discharged return stroke, and a higher area of negative charge. In a time less than about 100 ms, if the discharged return-stroke channel has additional charge available [1], a dart or continuous leader will pass the return-stroke channel and increase the ionization degree of the channel, then deposit more charges in the

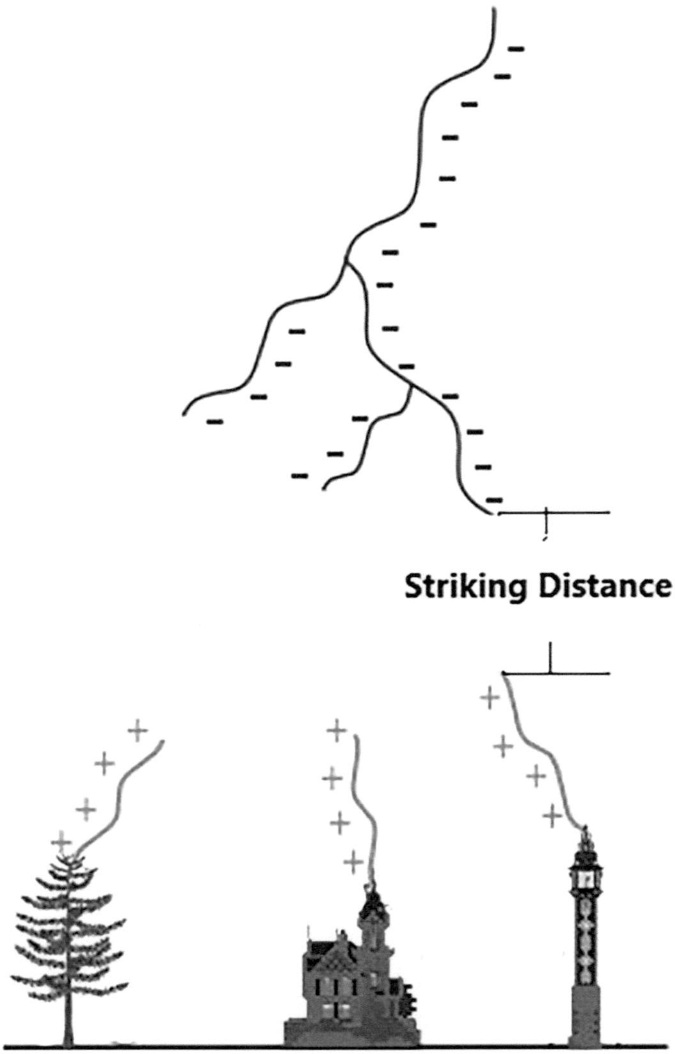

Fig. 1.13 The schematic description of striking distance

channel, and bring the potential of the cloud groundward once more. Therefore, the dart-type leader makes the second stage or subsequent return stroke. The luminous section of the channel is called the dart leader, and its length is about fifty meters [1]. It smoothly goes to the ground at about 2 × 106 m/s; the average value of the step leader velocity is less than this. A lightning flash can contain several strokes (on average, four strokes), and usually the first stroke's current amplitude is higher than the amplitude of the other strokes. Figure 1.15 shows the distribution of charges in a thundercloud.

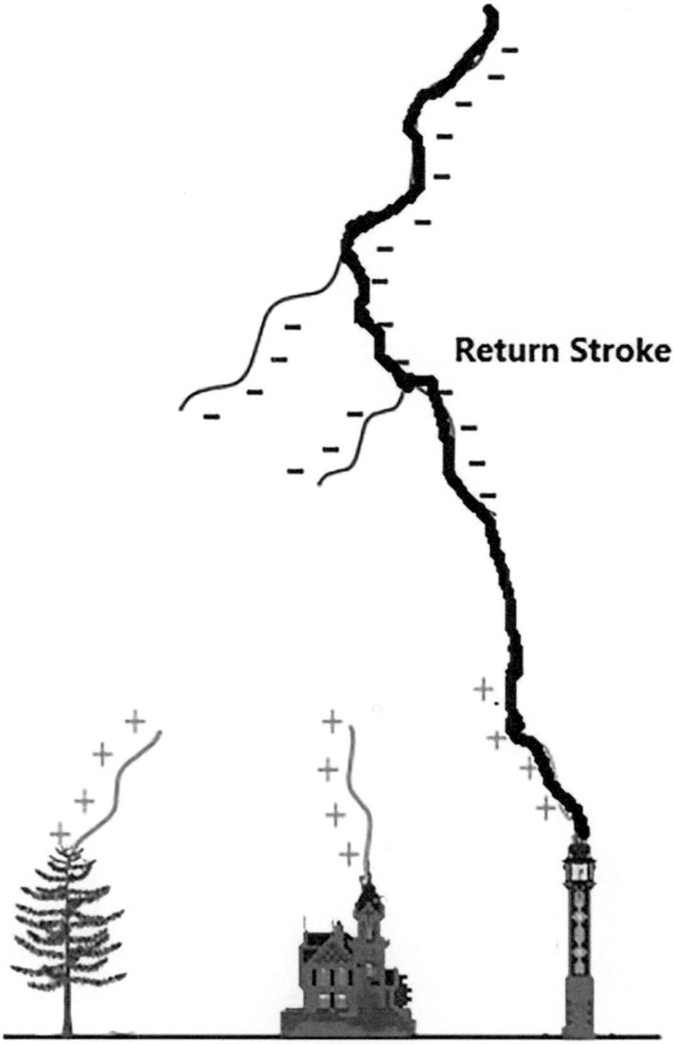

Fig. 1.14 The return stroke channel

Continuing Current

Sometimes, by reaching the return stroke channel to the cloud, the charge production in the thundercloud may continue, and the return stroke channel is fed for a long time with a lower current. This is named continuing current. The continuing current can be as long as several tens of milliseconds to several hundreds of milliseconds. This current amplitude range can be several tens of amperes to several hundreds of amperes [16]. Sometimes, the continuing current begins with subsequent strokes.

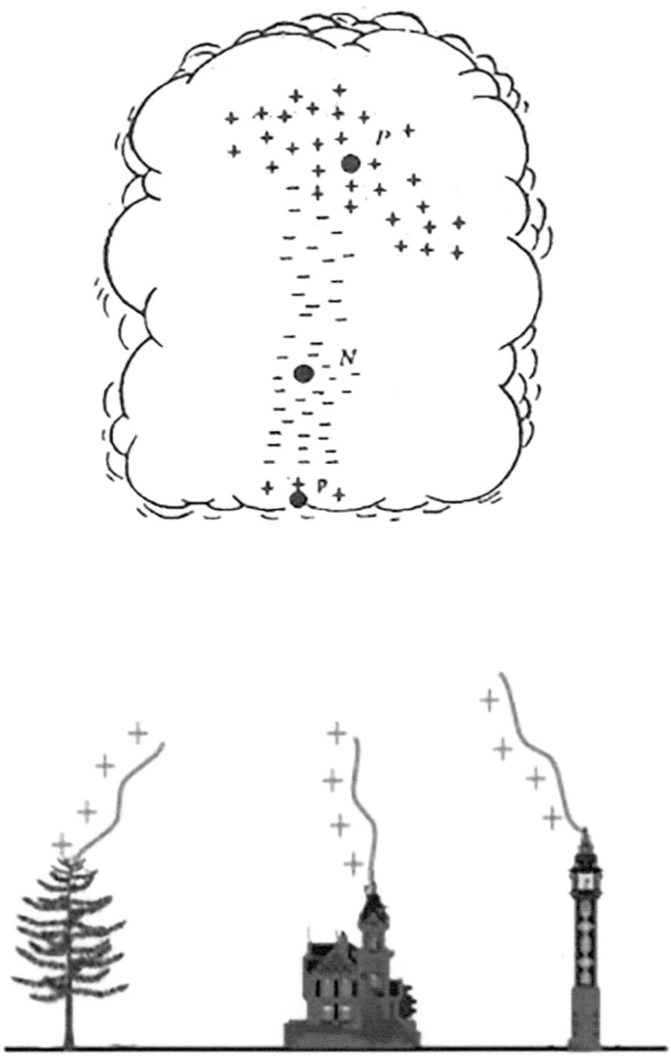

Fig. 1.15 Distribution of charges in a thundercloud

1.3 Importance of Protecting the Power System Against Lightning

As lightning strikes a power system, it can damage the power system apparatus, primarily electronic and electrical systems. The severity of this damage depends on the characteristics of the equipment and the lightning flash characteristics.

The transient wave is an overvoltage that lasts a few microseconds to a few milliseconds, with a duration less than a power frequency cycle [17]. Even if the amplitude of the transient does not exceed the standard operating voltage due to its high-frequency component, it can still be harmful. Lightning transient overvoltage is usually faster than other overvoltage.

A lightning strike can produce transient overvoltage on the power system in three ways:

1. Direct stroke to the phase conductor
2. Direct stroke to tower or shield wire
3. Stroke to the vicinity of the power system line.

The overvoltage is dependent on the lightning current amplitude, front time of the lightning current waveshape, return stroke channel velocity, and cloud height.

A lightning strike can damage the power system in two ways: (i) overvoltage on the system results in breakdown of insulation and shorts the power system. (ii) The energy of the stroke is much more than the energy that the system can tolerate; therefore, it can destroy the system equipment.

1.3.1 The Direct Stroke

Overhead transmissions have an extended length that passes through open areas; therefore, due to their long height, in flat areas, there is a high chance of being struck by direct lightning. A stroke to line can appear as a transient overvoltage that travels to both sides of the line. This overvoltage can break down the line insulation. Figure 1.16 shows a direct lightning stroke.

Many transmission lines have shield wires, but this shield wire misses the lightning strike, and lightning will reach the phase wire. For simulating the direct stroke to the line, the lightning current is simulated by a current source that is connected to the striking point. The lightning overvoltage on a line can be computed by simply following the formula:

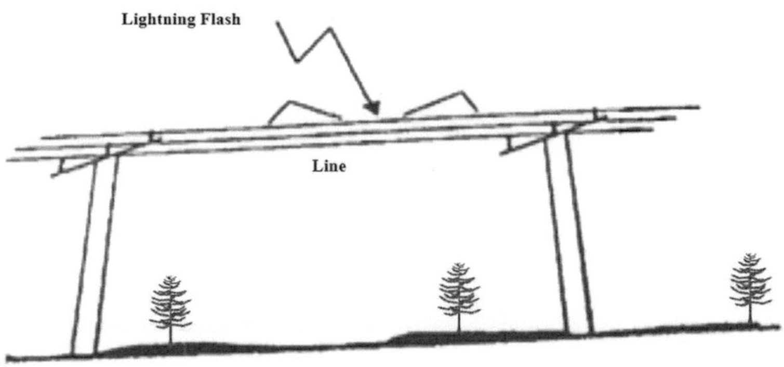

Fig. 1.16 Direct lightning strike to the line

$$V = \frac{I_p}{2} Z \tag{1.1}$$

I_p = Lightning current
Z = Surge impedance of line

1.3.2 Number of Strokes to the Line

The transmission line casts a shadow on the ground, which suggests that all flashes that terminate on this shadow will strike the transmission line [18, 19]. Figure 1.17 shows this shadow, whose width is W for a line with two shields. For lines with one shield, $d = 0$.

The H is called the mean shield wire height and can be computed from (1.2)

$$H = H_g - 2/3(H_g - H_{gw}) \tag{1.2}$$

where

H_g and H_{gw} are the height of the shield wire on the tower and at midspan, respectively.

The width of the shadow can be computed from the following formulas.

$$W = d + 4H \tag{1.3}$$

For better correlation with data, the following modification is applied [20].

$$W = d + 4H^{1.09} \tag{1.4}$$

where the units of d, H, and W are in meters, therefore, the number of strokes to the transmission line (N_L) is:

$$N_L = 0.012T(d + 4H^{1.09})\ \text{flashes/100km} - \text{yr} \tag{1.5}$$

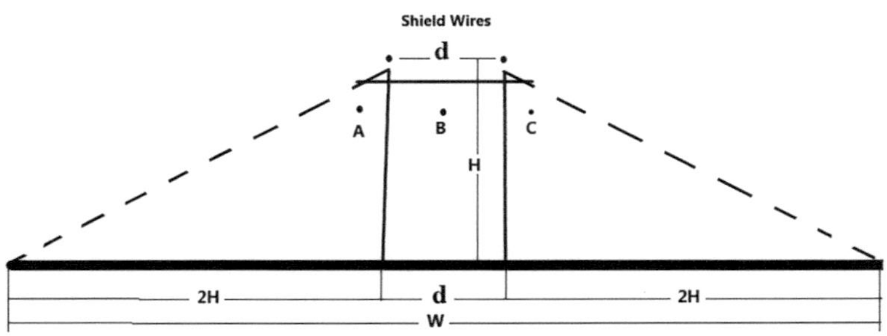

Fig. 1.17 The width of the shadow for the three-phase, double shield wire transmission line

where

T = Iso-keraunic level in the area where the line passes (thunder days/yr).

Many researchers believed that the number of flashes to the tower (N_T) is:

$$N_T = 0.6(N_L - N_S) \tag{1.6}$$

where

N_S = number of strokes to produce shielding failure per 100 km per year.

1.3.3 Strike Distance

Strike distance is a base parameter for the electro geometric model. Strike distance is the distance that when downward leaders reach this distance from the structure, the upward leader starts, and lightning strikes the structure. The structure that lightning enters first within its strike distance will be struck by lightning [21, 22], as can be seen in Fig. 1.18. As the downward leader enters the strike distance zone of the phase conductor, it will strike the phase conductor. If a downward leader enters the strike distance zones of shield wire or ground, the phase conductor will be protected.

As it is demonstrated in Fig. 1.18, the striking area for each wire is a circle that is centered on the wire and whose radius is the strike distance. If the downward leader is as "SW", lightning will touch the shield wire because it has entered the strike distance zone of the shield wire.

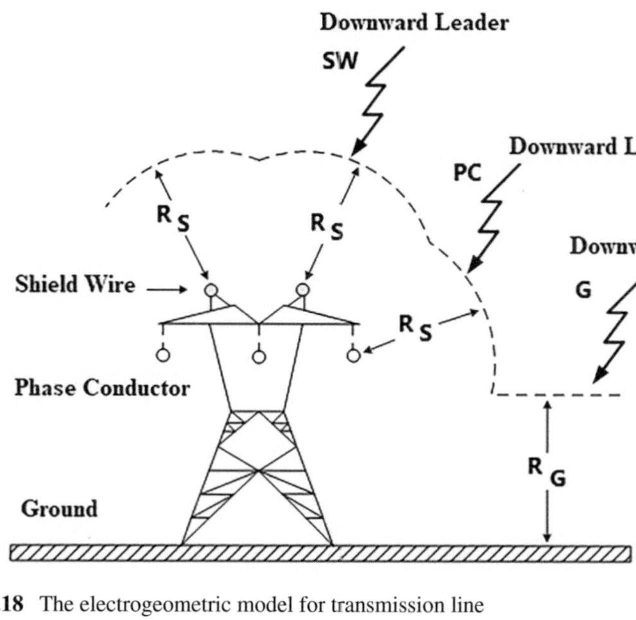

Fig. 1.18 The electrogeometric model for transmission line

If the downward leader is a "PC", lightning will touch the phase conductor, and in "G", it will touch ground.

The relation between the lightning current and strike distance is as follows:

$$R_S = KI^n \tag{1.7}$$

Different researchers introduced different values for K and n. For example, these parameters, which are suggested by Mousa, are $K = 8$ and $n = 0.65$ (I is the current amplitude of lightning in kiloampere and R_S is in meters) [23].

In some research, the same value is chosen for the strike distance of the phase conductor, shield wire, and ground, but many researchers have suggested different values for the ground strike distance.

$$R_G = K_G R_S \tag{1.8}$$

K_G = ratio of ground strike distance to strike distance of phase conductor (or shield wire). Erikson's suggestions for K_G are as follows [22, 24]:

$K_G = 1$ for HV transmission lines

$K_G = 0.8$ for EHV transmission lines

$K_G = 0.67$ for UHV transmission lines

The following values are suggested for K_G by IEEE Std. 1243 [25].

$K_G = 0.55$ ($H \geq 40$ m)

$K_G = 0.36 + 0.17\ln(43\text{-}H)$ ($H < 40$ m)

1.3.4 *Using the Electrogeometric Method to Calculate the Unprotected Area*

An unprotected area of transmission line is shown in Fig. 1.19.

$$X_{UP} = \text{PL}-\text{PW} \tag{1.9}$$

$$\text{PE} = R_S \cos\theta \tag{1.10}$$

where X_{UP} is unprotected width.

$$\theta = \arcsin\left(\frac{K_G R_S - Z_P}{R_S}\right) \tag{1.11}$$

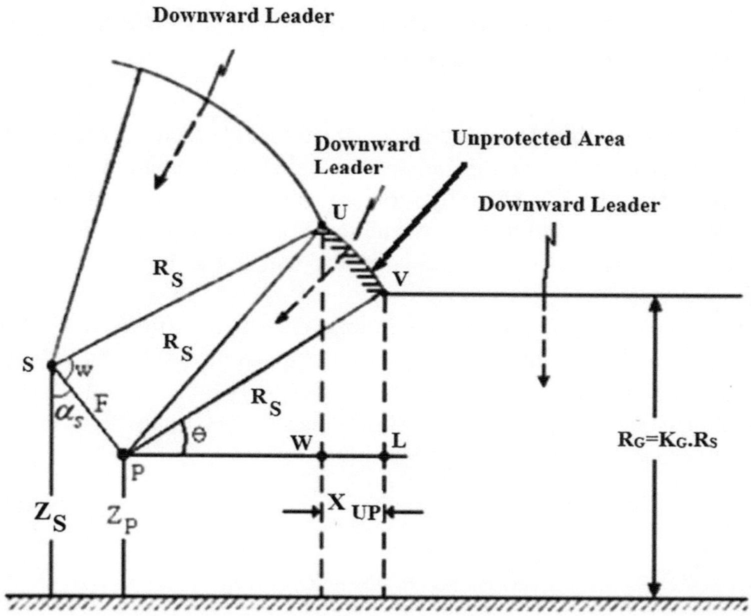

Fig. 1.19 Electro geometric model for the calculation of the unprotected area

PW can be calculated as:

$$\mathrm{PW} = R_\mathrm{S} \sin(w - \alpha_\mathrm{s}) \tag{1.12}$$

$$w = \arcsin\left(\frac{F}{2R_\mathrm{S}}\right) \tag{1.13}$$

$$\alpha_\mathrm{s} = \arctan\left(\frac{X_\mathrm{P} - X_\mathrm{S}}{Z_\mathrm{S} - Z_\mathrm{P}}\right) \tag{1.14}$$

$$X_\mathrm{UP} = R_\mathrm{S}[\cos\theta + \sin(\alpha_\mathrm{s} - w)] \tag{1.15}$$

If $K_\mathrm{G} R_\mathrm{S} < Z_\mathrm{P}$ then $\cos\theta = 1$, therefore:

$$X_\mathrm{UP} = R_\mathrm{S}[1 + \sin(\alpha_\mathrm{s} - w)] \tag{1.16}$$

1.3.5 Using the Electrogeometric Method to Calculate the Effecting Shielding Angle

Generally, when lightning stroke to phase conductor is not produce flashover between the phase and ground the shielding angle is an effecting shielding. In order to have effecting shielding the X_{UP} should go to zero. As Lightning current increases the strike distance increase and the X_{UP} decreases. For an individual lightning current amplitude X_{UP} is zero, and effecting shielding will occur. Figure 1.20 shows an effecting protection.

Assume that I_{max} is amplitude of lightning current which effecting protection occurs. Lightning current more than I_{max} cannot touch the phase and will strike to the shield wire or earth. The effecting shielding angle is shown in Fig. 1.20 (α_E).

α_E will be calculated for two outer phases because inner phase will be shielded by outer phases and shield wire.

If $X_{UP} = 0$.

Then;

$$X_S = SN - PR \tag{1.17}$$

$$X_S = \sqrt{R_S^2 - (K_G R_S - Z_P)^2} - \sqrt{R_S^2 - (K_G R_S - Z_S)^2} \tag{1.18}$$

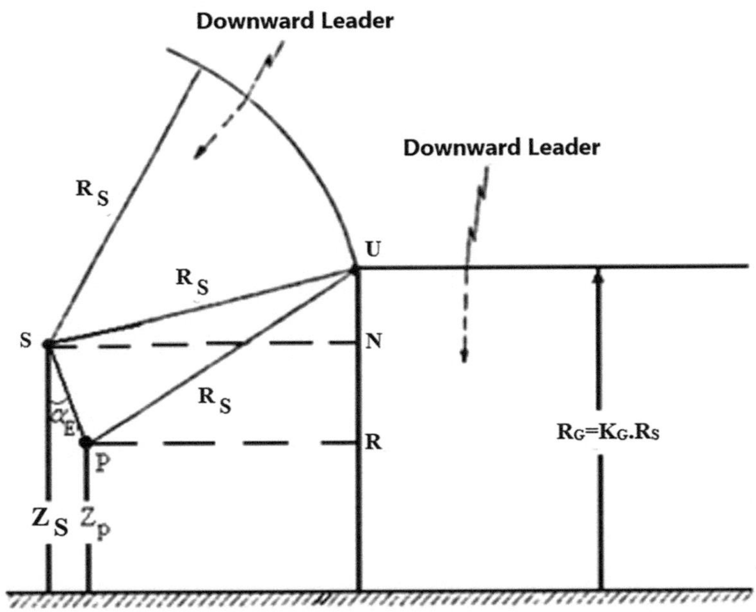

Fig. 1.20 An effecting shielding

Therefore, the effective protection angle is:

$$\alpha_E = \arctan\left(\frac{X_S}{Z_P - Z_S}\right) \tag{1.19}$$

1.3.6 Shielding Failure Rate (SFR) Computation

The number of strokes per phase per 100 km-year is called the shielding failure rate. I_{min} is the minimum lightning current amplitude that can strike the phase wire. $I_{min} = 3\ kA$ is selected by Hileman [26], which was recorded by CIGRE.

The unprotected width equals X_{UP} for $I = I_{min}$, and for $I = I_{max}$, the unprotected width equals zero; therefore, the average value of the unprotected width is $X_{UP}/2$.

If the number of lightning strikes to ground per km^2-year is γ, so the number of strokes to the unprotected width is:

$$N_{\text{Unprotected Width}} = 0.1\gamma\frac{X_{UP}}{2} \tag{1.20}$$

γ it is obtained from the weather bureau of the area, but usually, flash ground density is not available, and γ it is calculated from the isokerunic level of the area. Eriksson [27] introduced the following formula for γ:

$$\gamma = 0.12T \tag{1.21}$$

T is the number of thunderstorm days per year.

Therefore;

$$N_{\text{Unprotected Width}} = 0.012T\frac{X_{UP}}{2} \tag{1.22}$$

SFR occurs for currents in the range of (I_{min} to I_{max}), then:

$$\text{SFR} = 0.012T\frac{X_{UP}}{2}(P_{I_{min}} - P_{I_{max}}) \tag{1.23}$$

P_I is the probability that the amplitude of lightning current exceeds, and is obtained from data from different research in this field. The following formulas are introduced by Popolansky (1.24) and Anderson-Erikson (1.25) [18].

$$P_I = \frac{1}{1 + \left(\frac{I}{25}\right)^2} \tag{1.24}$$

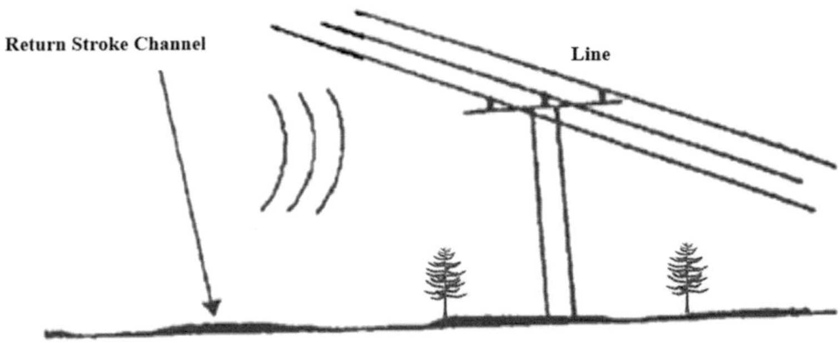

Fig. 1.21 Schematic representation of the induced voltage process on a line

$$P_{\mathrm{I}} = \frac{1}{1 + \left(\frac{I}{31}\right)^{2.6}} \tag{1.25}$$

1.3.7 Indirect Stroke

When lightning strikes near a line, the electromagnetic field of the flash can induce a high transient voltage on the line that can be harmful for low-voltage and medium-voltage transmission lines. Figure 1.21 shows a schematic representation of this process.

The induced voltage on the distribution system line can transfer to the LV side of the distribution transformers and could damage the LV equipment. The shield wire that is used for the protection of the phase conductor against direct strike can reduce the induced voltage as well.

The following parameters influence the amplitude of the induced voltage on the line [28]:

(i) Line height
(ii) Distance between the striking point and the line
(iii) Amplitude and front time of the lightning current wave
(iv) The velocity of the return stroke wave
(v) Cloud height.

As shown in Fig. 1.22, if the distance between the leader and the line is more than the critical value (y_1), the downward leader will touch the earth; otherwise, it strikes the overhead line.

where

$$y_1 = \sqrt{R_S^2 - (R_S - H_P)^2} \quad \text{for } R_S > H_P \tag{1.26a}$$

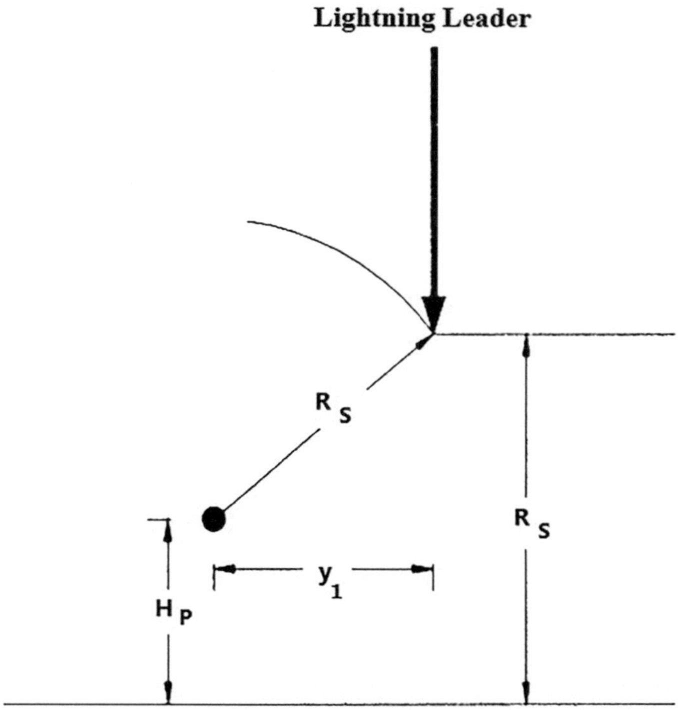

Fig. 1.22 Schematic diagram for the computation of the critical distance

$$y_1 = H_P \quad \text{for}\, R_S \leq H_P \tag{1.26b}$$

If lightning does not touch the overhead line, then it will strike near the overhead line, and the induced voltage on the overhead line can be higher than the BIL of the power distribution lines.

References

1. M.A. Uman, *Lightning* (McGraw-Hill Inc., U.S.A., 1969)
2. D. Baskin, Lightning without cloud. Bul. Am. Meteor. Soc. **33**, 348 (1952)
3. H.F. Gisborne, Lightning from a clear sky. Monthly Weather Rev. **56**, 108 (1928)
4. F. Myers, Lightning from a clear sky. Monthly Weather Rev. **59**, 39–40 (1931)
5. R.H. Golde (Ed.), Chapter 3, *Lightning Vol. 1: Physics of Lightning* (Academic Press Inc., 1977)
6. B. Vonnegut, Some facts and speculations concerning the origin and rate of thunderstorm electricity. Met. Monogr. **5**, 224–241 (1963)
7. B. Vonnegut, Thundercloud electricity. Discovery **26**, 12–17 (1965)
8. B.J. Masson, *The Physics of Clouds*, 2nd edn (Clarendon Press, Oxford, 1971), p. 671
9. V. Cooray (Ed.), *The Lightning Flash* (IET, UK, 2008)

10. V. Cooray (Ed.), *Lightning Electromagnetics* (IET, UK, 2012)
11. V.A. Rakov, M.A. Uman, *Lightning Physics and Effects* (Cambridge University Press, UK, 2003)
12. G. Grenet, Essai d'explication de la charge électrique des nuages d'orages. Extrait annls Geophys. **3**, 306–307 (1947)
13. B. Vonnegut, Possible mechanism for the formation of thunderstorm electricity. Bull. AM. Met. Soc. **34**, 378 (1953)
14. R.H. Golde (Ed.), *Chapter 5, Lightning Vol. 1: Physics of Lightning* (Academic Press Inc., 1977)
15. K. Berger, Development and properties of positive lightning flashes at Mount Salvatore, in *Culhm Conf., Section I2* (1975)
16. V. Cooray, *An introduction to Lightning* (Springer, UK, 2015)
17. V. Cooray (Ed.), *Lightning Protection* (IET, UK, 2010)
18. J.J. LaForest (Ed.), Transmission line reference book 345 kV and above, Chapter 12, in *Lightning Performance of Transmission Lines* (EPRI, California, USA, 1982)
19. R. Lings (Ed.) *EPRI Transmission Line Reference Book 200 kV and Above*, 3rd edn (EPRI, California, USA, 2005)
20. R.H. Golde (ed.), *Lightning vol. 2, Protection of Transmission Lines* (Academic Press, NY, USA, 1977), pp. 697–745
21. R. Golde, The frequency of occurrence and the distribution of lightning flashes to transmission lines. Electr. Eng. **64**, 902–910 (1945)
22. F.S. Young, J.M. Clayton, A.R. Hileman, Shielding of transmission lines. IEEE Trans. Power App. Syst. **S82**, 132–154 (1963)
23. A.M. Mousa, The applicability of lightning elimination devices to substations and power lines. IEEE Trans. Power Delivery **13**, 1120–1127 (1998)
24. G.W. Brown, E.R. Whitehead, Field and analytical studies of transmission line shielding: part II. IEEE Trans. Power App. Syst. **PAS-88**, 617–626 (1969)
25. IEEE Guide for Improving the Lightning Performance of Transmission Lines, IEEE Std. 1243, (1997)
26. A.R. Hileman, in *Insulation Coordination for Power Systems* (CRC Press, 1999)
27. A.J. Eriksson, The incident of lightning strikes to power lines. IEEE Trans. Power Delivery **2**, 859–870 (1987)
28. P. Chowdhuri, *Electromagnetic Transients in Power Systems*, 2nd edn. (Prentice Hall, India, 2008)

Chapter 2
The Effect of Lightning on the Power Distribution Networks

Abstract Generally, computation of overvoltage due to indirect lightning stroke is complicated than overvoltage due to direct lightning stroke. In this book indirect induced overvoltage on distribution system are analyzed, therefore in this chapter effects of indirect induced overvoltage on power distribution networks are analyzed and different types of surge arresters are introduced.

2.1 Introduction

In different regions of the world, lightning can happen, but in some regions, the lightning activity is more intense than in other regions, such as South America, the Central United States of America, and some regions in Africa. In these regions, there are proper meteorological conditions such as warm temperatures and proper solar radiation, which lead to the production of thunderstorm clouds [1, 2].

Generally, computation of overvoltage due to indirect lightning stroke is complicated than overvoltage due to direct lightning stroke and needs the following steps.

(i) Return stroke channel modeling
(ii) Computation of electromagnetic fields near the line
(iii) Computation of induced voltage due to coupling between electromagnetic field and line conductors.

Overvoltage on the power distribution due to direct stroke or indirect stroke (which is in the scope of this book) and transfer to the low-voltage side of transformers can have harmful effects [2, 3].

B. Vahidi and A. F. Nematolahi, *Lightning Induced Overvoltage and Current on Distribution Power Systems*, https://doi.org/10.1007/978-981-95-3712-9_2

2.2 Lightning Strikes on Transmission Lines

As is known, power transmission lines have a long trace and high height; therefore, they are exposed to lightning strikes, and when a lightning strike directly hits a transmission line, events such as the following can happen [4–7]:

1. **Overvoltage on the Transmission Lines**

Overvoltage due to lightning can initiate a transient overvoltage on the lines. Lightning strikes can cause transient overvoltage on the transmission lines, which can exceed the insulation level of the lines and equipment and may damage the insulation.

2. **Power System Transients**

Lightning may produce fast transient current on the power system, which travels through lines and could generate high-frequency harmonics and voltage spikes on the power networks. Sensitive apparatus can be disrupted and may produce a fault, a malfunction, or push the protective devices to operate.

3. **Damage to Transformers and Substations**

Induced high current and voltage due to a lightning strike near transmission lines, which can cause damage to the insulation of the winding, and the transformer can be damaged. The other equipment of the substation, like circuit breakers, insulators, and capacitors, can be damaged as well.

4. **Disruption in the Grid**

As the lightning discharge energy is high, the lightning discharge can cause temporary disruption in the grid network due to tripping the protection devices and produce a power interruption, which can affect the industries and other customers of the electric utility.

5. **Communication Systems**

Communication systems in the vicinity of transmission lines can be affected due to the electromagnetic field of a lightning strike, which can affect the operation of systems such as the communication system, control signals, data transmission, and monitoring systems.

6. **Safety Risk**

Safety risk to workers who are working in the vicinity of transmission lines during lightning activity is another impact of lightning strikes to transmission lines.

2.3 Effects of Direct and Indirect Lightning Strokes to Distribution Lines

Direct and indirect lightning strokes can cause different impacts on distribution lines and components of the distribution system:

1. **Components Damage**

Components (conductors, insulators, transformers, …) of the distribution line can be damaged by lightning strikes. The damage to these components can cause a power distribution outage and will require repair or replacement of the affected components.

2. **Surges**

Induced surges due to lightning strikes can damage the components' insulation and trip the protective devices, which can disrupt the power supply to customers.

3. **Reduction of Power Quality**

Power quality, which is delivered to customers, can be reduced due to the lightning strikes to the line, because lightning flashes can produce flicker, voltage sag, and interruption in the power supply process. These fluctuations can disrupt sensitive equipment function or critical processes.

4. **Grounding Effects**

The grounding system of the distribution network can be affected by lightning strikes. As a high current can be induced during the lightning strike process, the grounding potential of the distribution grounding system can rise sharply, which can endanger workers' safety or damage the devices.

5. **Fire Risk**

Sometimes, a lightning strike on the distribution system can produce fire due to high current, which can produce high temperatures, and in the presence of ignitable materials, the fire risk increases.

6. **Malfunctiong**

Overvoltage can be harmful to power system equipment. Overvoltage can cause inefficient illumination of lights due to overheating and burnout of bulbs. Overvoltage can reduce the efficiency of electrical machinery due to overheating and damage to insulation.

7. **Hazard to the PCBs**

High voltage can destroy circuit boards and could cause a fire.

8. **Sensitive Load Trip**

Some sensitive loads, such as UPS, could trip or be damaged.

9. **Overheating**

Electronic components can be damaged due to overheating.

10. **Higher Power Consumption**

Due to overvoltage, an electrical device may draw higher current, which means higher energy consumption.

11. **Conductor Internal Damage**

Overvoltage may result in electromigration, which can reduce the reliability of the circuit and produce loss of connections.

12. **Short Circuit**

Overvoltage can produce a flash, which may produce a short circuit in the system.

13. **Decreasing Lifetime**

Overvoltage can degrade the insulation of the component and reduce its lifetime, which may cause permanent failure.

14. **Increasing Starting Current of Motors**

Overvoltage can increase the starting current and complete load current as well. At 110% of nominal voltage, the starting current will increase 10–12% which can put pressure on the power supply.

The above-mentioned damages to the power distribution system due to lightning have forced engineers to protect the distribution system against direct and indirect lightning overvoltage. Due to the fact that distribution lines are passing near structures or trees, which helps to protect the distribution lines against direct strikes, computing the induced voltage on distribution lines due to indirect lightning strikes and protecting the line against it is very important.

2.4 Surge Protection

One of the most common transients in the electrical distribution networks is surges due to indirect lightning strikes [3]. This book concentrates on lightning-induced voltage surges in distribution systems.

Two important parameters of the field in induced voltage calculation are the peak and rate of rise of the field, which they have influence the peak of the induced voltage [3].

The peak of induced voltage on metallic surfaces due to the lightning field is related to the peak of the electric field, but the peak of voltage produced in a wire loop is related to the rate of change of the magnetic field [3].

2.4.1 Surge Protection Devices

Surge protection devices (SPDs) are designed to conduct under surge conditions. The SPDs are installed in each phase of three-phase systems. In normal conditions, an SPD is doing nothing like a fuse; when passing current, it is in its normal range. But when the voltage on the network rises due to lightning, the surge protection devices

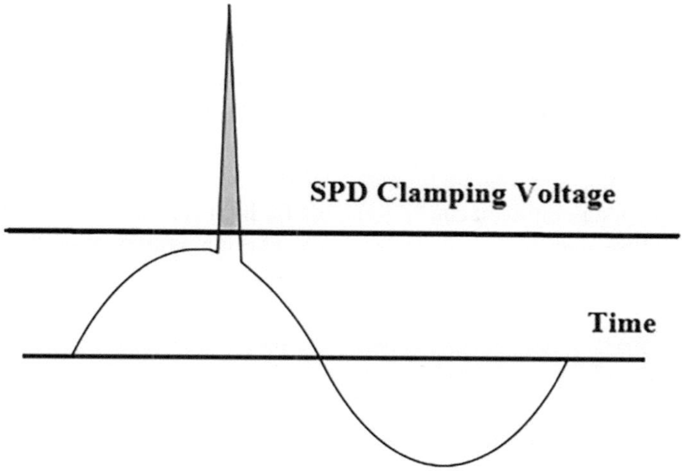

Fig. 2.1 Clamping voltage of SPD

start to conduct and divert the energy away from the equipment. The overvoltage lasts for a short time (microseconds to milliseconds). Unlike fuses, the SPDs can be worked on many times [8].

When the voltage on SPDs rises above a specific value, these devices are triggered. This particular value is known as clamp voltage [8]. The clamp voltage should be higher than the peak of the standard voltage plus a margin in order to prevent SPD conduction during the normal operation of the system. Figure 2.1 shows a surge voltage imposed on the normal voltage.

As can be seen, the shaded region of the wave shows that SPD clamps the surge and prevents damage to equipment. The SPD's current capacity should meet the amount of surge current.

The SPD technology should have the following criteria:

1. Can handle the surge in applications.
2. The clamping voltage is high enough to avoid oration in normal conditions.
3. Switching off after passing the surge.

SPDs are usually made of a metal oxide varistor or a spark gap. Metal oxide surge arresters and spark gaps have a high-current capacity and different voltage ratings. Still, spark gaps, when operating, effectively shorten the supply, and they need a special design to clear the short condition.

Some SPDs, like gas discharge tubes, should not be connected across the mains because they may never turn off, although some spark gaps will extinguish themselves. MOVs with different voltage ratings will turn off after passing the surge to ground and can be used for all installations.

MOVs, which are used in surge arrester devices, should do their job for many years. When these MOVs are used in high lightning activity areas, their lifetime can be reduced. If the MOVs reach their end of life, their impedance drops, and they draw a high amount of current. Some standards recommend using an external current limiting device together with SPDs [8].

Those fuses that are used with surge protection devices are not for the protection of SPDs; they are used to protect the mains from the failure of SPDs due to high-current surges. Small fuses for protecting the SPD would fail during large surges; therefore, fuses that can tolerate high-current surges could be used. If the supply's own fuse has enough rating, it can be used, but for high current supplies, installation of a separate fuse spur for devices is necessary. This configuration has the benefit of allowing the continued supply of current to the installation.

2.4.1.1 Types of Surge Arrester

The selection of surge arresters depends on factors such as [9]:

i. Voltage
ii. Current
iii. Reliability
iv. Space available for installation.

Different types of surge arresters are as follows:

- Rod Gap Arrester
- Sphere Gap Arrester
- Horn Gap Arrester
- Multiple-Gap Arrester
- Impulse Protective Gap Arrester
- Electrolytic Arrester
- Expulsion Type Lightning Arrester
- Valve Type Lightning Arresters
- Thyrite Lightning Arrester
- Auto Valve Arrester
- Oxide Film Arrester
- Metal Oxide Lightning Arrester.

1. **Rod Gap Arrester**

A simple way to protect electrical and electronic devices from surges due to lightning and other high voltage surges is to use a rod gap arrester [10, 11]. The rod gap arrester is made of two rods that are placed a small distance apart, with one of them connected to the line and the other to ground. Figure 2.2 shows the protection of an insulator string by a rod gap arrester.

When a surge, such as lightning overvoltage, comes on a rod gap arrester, the overvoltage across the gap makes it spark, and the lightning current is directed to ground, protecting the devices from damage. One of the limitations of rod gap application is the continuation of the arc between rods, which can damage the rods or pose a hazard to system equipment. In order to prevent the damage, a series of

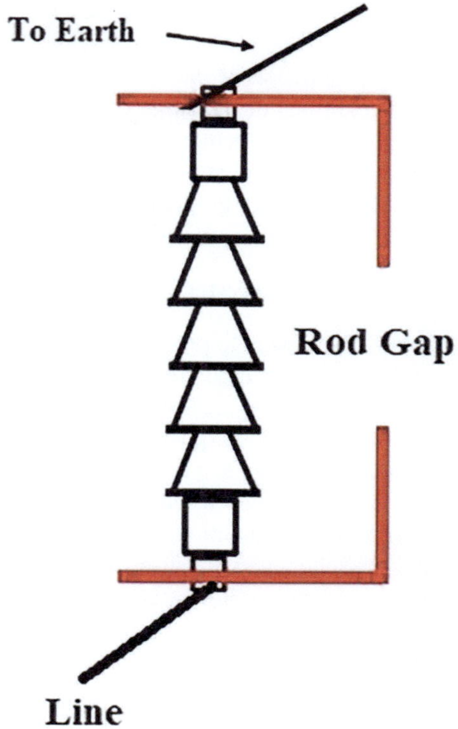

Fig. 2.2 Schematic presentation of the rod gap arrester

current-limiting reactors can be used with the rod gap. To prevent damage to system equipment, the gap setting of the arrester must be carefully chosen to break before the equipment is damaged [11].

Advantages and Disadvantages of Rod Gap Type Arresters:

i. Advantages:

   - Simple configuration, installation, and maintenance of rod gap type arresters are relatively inexpensive.
   - Rod gap arresters are effective in conducting the lightning current to the ground, which can protect the power system from lightning. It is possible to use the rod gap arresters on a variety of structures, such as buildings, towers, and poles.

ii. Disadvantages:

   - Following the power frequency current after passing the surge current to ground, the arc between rods will not be extinguished after the initial surge current passes to ground. Power frequency current may still pass to the ground, which can cause damage to power system equipment and can trigger the circuit breaker, which interrupts the power supply to the system.

- Due to the spark between rods, a high amount of heat is produced, which can cause damage to the rods (weakened or broken).
- The function of the rod gap surge arrester can be affected by atmospheric conditions or the surge polarity. Atmospheric conditions such as humidity, pressure, and temperature can influence performance.
- Due to the limitations of the rod gap arresters, they are used as backup protection, and other types of lightning protection, such as metal oxide arresters, are used as primary protection.

2. **Sphere Gap Arrester**

Another simple way to protect electrical and electronic devices from surges due to lightning and other high voltage surges is to use a sphere gap arrester [10, 11]. The sphere gap arrester is made of two spheres that are placed a small distance apart from each other. One of the spheres is connected to the line and the other to the ground. The choke coil (Fig. 2.3) is used to limit the current.

The distance between spheres is adjusted so that the sphere gap is not firing under normal operating conditions [11]. However, if a high voltage surge comes on the line across the sphere gap, it causes the sphere gap to fire and conduct the surge current to the ground, protecting the equipment from damage. The choke coil is used to limit the current.

Advantages and Disadvantages of sphere gap type arresters:

i. Advantages:

- This kind of arrester is low cost.
- The impulse ratio of the sphere gap arrester is unity and, therefore, can withstand the surge overvoltage equal to the magnitude of the system's rated voltage. It can also be used as a reliable protection in electrical networks [11].

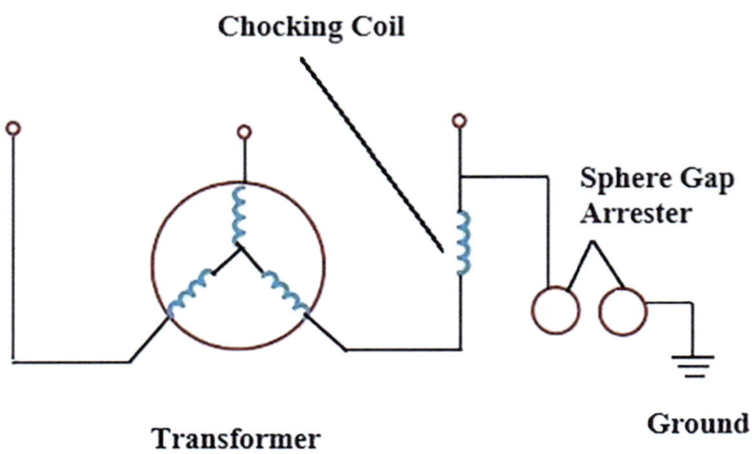

Fig. 2.3 Schematic presentation of the sphere gap arrester

ii. Disadvantages:

After the initial surge, the current passes to ground, and the arc between spheres will not be extinguished. Power frequency current may still pass to the ground, which can cause damage to power system equipment and can trigger the circuit breaker, which interrupts the power supply to the system; therefore, it may not be desirable in some applications.

3. **Horn Gap Arrester**

A simple way to protect electrical and electronic devices from surges due to lightning and other high voltage surges is to use a horn gap arrester [10, 11]. The horn gap arrester is made of two horns, which are placed a small distance apart from each other. The gap between the horns is smaller at the bottom and larger at the top. One of the horns is connected to the line and the other to the ground. During operation under normal conditions, the gap does not break because the strength of the gap is higher than the standard operating voltage. But at a surge voltage higher than the strength of the gap, it breaks, and the current is conducted to ground. The air around the arc is heated, and the electromagnetic field forces the arc to travel upward. When the gap becomes wide enough at a certain point, the voltage is insufficient to maintain the arc; therefore, in a few seconds, the arc is extinguished [11].

The resistor in series with the gap reduces the passing current, and the choking coil will reduce the sharpness of the incident wave. Figure 2.4 shows a schematic representation of a horn gap arrester.

i. Advantages:

- Extinguishing an arc by itself without the help of any external equipment (such as a circuit breaker), which can prevent the interruption of the electrical supply.

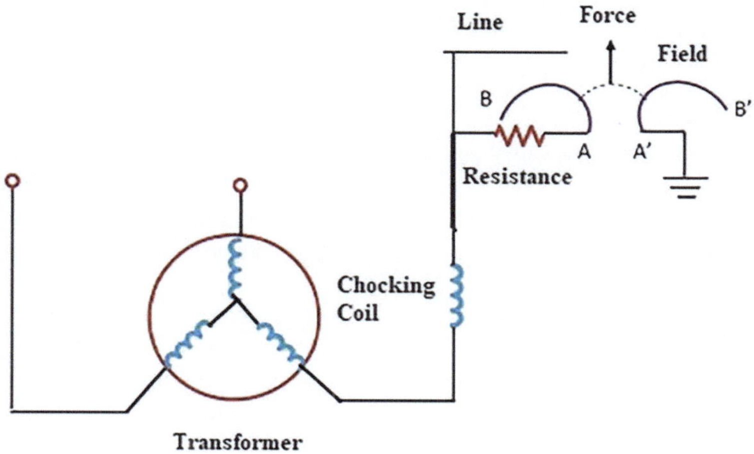

Fig. 2.4 Schematic presentation of the horn gap arrester

- Reduction in passing current through the arrester due to the series resistor, which can reduce the risk of damage to equipment.

ii. Disadvantages:

- A horn gap arrester can be shorted by external objects such as trees or a piece of wire, which can increase the risk of damage to the network equipment.
- Pollution and corrosion can affect the setting of the horn gap arrester, which can increase the risk of damage to the network equipment.
- The operation time of the horn gap arrester is around 3 s, which, in some applications where a fast response is required to protect the network, is not a proper device.

4. **Multi Gap Lightning Arrester**

A multi-gap arrester is made of several metallic cylinders, which are made from zinc alloy, and they are insulated from each other [12, 13]. The cylinders are separated by small air gaps (Fig. 2.5).

Cylinder A is connected to the line, and cylinder C is connected to the ground through a series resistance. The voltage of the line dictated the required number of gaps. The series resistance limits the arc, but it reduces the protection against surges, and in order to overcome this difficulty, the shunt resistance is added to some of the gaps.

During normal conditions, the voltage between the A and C cylinders is not enough to break down the series gap. But when a surge comes on the arrester, breakdown occurs between cylinders A and B, and a large current passes through the shunt resistance. After the surge passes, there is no arc between the B and C cylinders, and the remaining power passes through the shunt and series resistances to the ground [12, 13].

The series and shunt resistances are limiting the remaining power. Therefore, a small value of current is left, which is not enough to establish the arc between the A and B cylinders, and the system returns to its normal operation condition. This kind of arrester is custom-made to use in power systems up to 33 kV.

i. Advantages:

- The multi-gap arrester has a simple construction and is easy to install and maintain.
- Arc in a multi-gap arrester does not persist, and after passing the surge to ground, the arc extinguishes under normal voltage, which prevents damage to equipment.

ii. Disadvantages/Limitations:

- The multi-gap arrester is more expensive than the rod-type arrester.
- Proper adjustment of the gap for proper functioning.
- Typically, it can be used in systems up to 33 kV.

5. **Impulse Protective Gap Arrester**

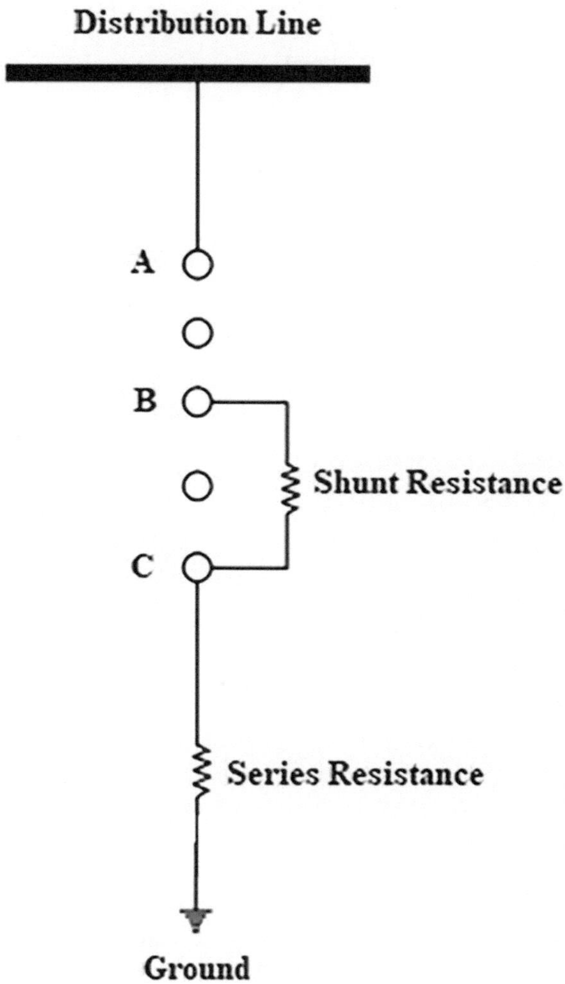

Fig. 2.5 Schematic presentation of the multi-gap arrester

The impulse voltage ratio of the impulse protective gap arrester is low, even less than one [11–13], and capable of extinguishing the arc. The method of impulse protective gap arrester working is as follows, and its schematic presentation is shown in Fig. 2.6.

This arrester consists of G_1 and G_2 electrodes, which are respectively connected to the line and ground, and needle N is in the middle of them.

At system frequency, the impedance of C_1 and C_2 is equal and large enough compared to the impedance of the resistor. Therefore, the potential of N (midway between G_1 and G_2) is not affected by flashover between G_1 and G_2.

During the transient, the impedance of C_1 and C_2 decreases, and the impedance of the resistor is now effective. Now, the whole of the voltage appears between N and G_1. The gap between N and G_1 breaks down suddenly, and the gap between N and G_2 immediately follows.

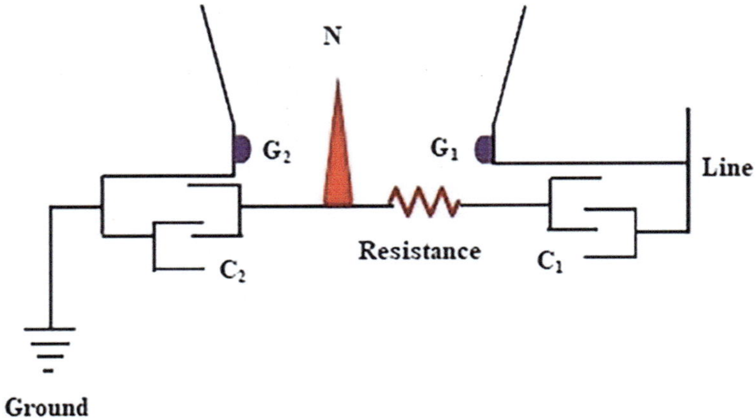

Fig. 2.6 Schematic presentation of the impulse protective gap arrester

Due to the unity impulse ratio of the sphere gap, it is working properly against surge voltages, but the sphere gap cannot act as an arc self-extinguishing device. In order to extinguish itself, the horns are added (Fig. 2.6).

6. **Electrolytic Arrester**

The electrolytic arrester operation is based on using a thin film of aluminum hydroxide, which is deposited on aluminum plates immersed in electrolyte, and behaves like a high resistor under low voltage but as a low resistor under a voltage higher than a critical value, and conducts the current to the earth to protect the devices. In order to have a higher critical voltage value, several films should be added on top of each other [11–13]. The typical value for the essential voltage of an electronic arrester is about 400 V. The resistance comes back to its high value after reducing the voltage to its normal value, and it blocks the current flow to the earth.

Usually, this kind of arrester is used in conjunction with an impulse gap arrester, which is designed to have a low impulse voltage ratio and is capable of self-extinguishing the arc. Figure 2.7 shows this configuration.

i. Advantages:

- Electrolytic arrester has a high-current discharge capacity.
- An electrolytic arrester does not need a series of resistors to reduce the flow current.

ii. Disadvantages/limitations:

- The formation of film on the electrodes has an influence on the operation of the electrolytic surge arrester. If the formation of the film is not proper, the arrester does not operate properly and needs frequent maintenance, which can increase the total cost of the arrester.

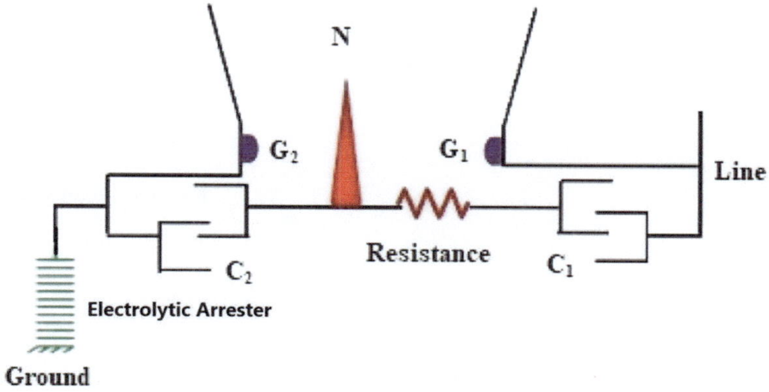

Fig. 2.7 Schematic presentation of the combination of the impulse protective gap arrester and the electrolytic arrester

- During cold weather times, the electrolytic solution can be frozen, making the arrester less effective or unusable.

7. **Expulsion Type Lightning Arrester**

The expulsion-type lightning arrester is known by different names; these are the protector tube or expulsion gap. The expulsion-type lightning arrester can be used to protect equipment against lightning surges [11, 13]. This arrester is typically used in power systems up to 33 kV. Figure 2.8 shows an expulsion-type arrester.

The expulsion arrester consists of an isolating interrupting spark gap (internal gap), which is in series with an external rod gap. The internal gap is the interrupting spark gap, which is enclosed in a fibrous tube.

The upper electrode of the internal gap is connected to the lower electrode of the rod gap, while the lower electrode of the internal gap is connected to the ground. Each phase of the line is connected to an arrester.

When a surge comes on the expulsion arrester, the rod gap sparks, and an arc is produced between the electrodes of the internal gap. The fiber of the tube walls vaporizes, and in a short time, a high-pressure gas is made inside the tube. The high-pressure gas expels the ionized air from the tube, and the arc is extinguished when the current passes zero because of the strong deionizing effect. After extinguishing the arc, the arrester is ready for the next operation.

i. Advantages:

- An expulsion-type arrester is relatively cheaper than some other types of arresters.
- A power frequency following current does not pass the arrester after the initial surge passes, which reduces the risk of damage to the system.
- Expulsion-type arrester installs easily.

ii. Disadvantages/Limitations:

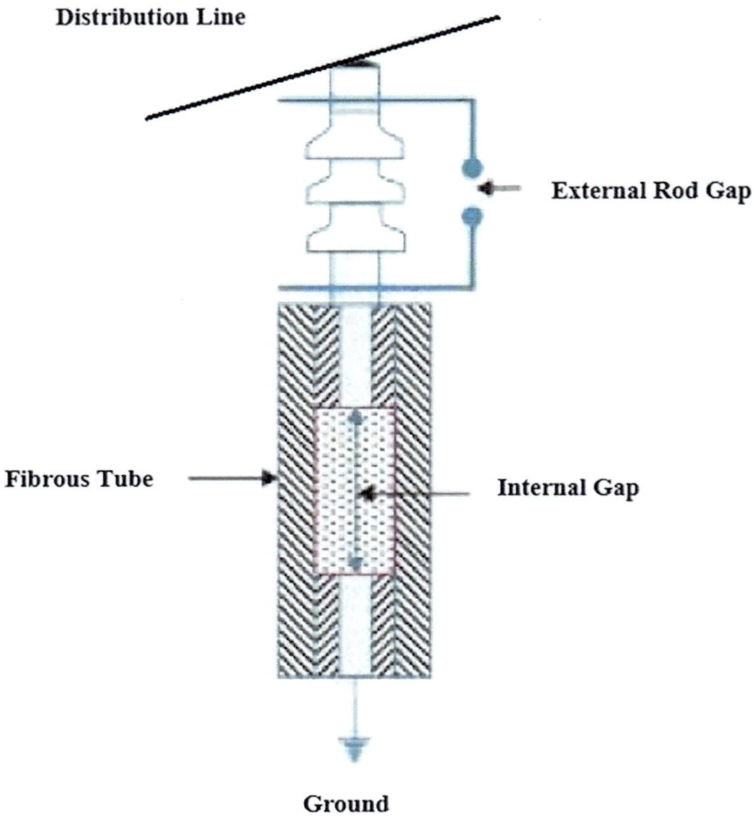

Fig. 2.8 The expulsion-type arrester

- Due to the consumption of tube material, this kind of arrester has a limited lifespan.
- Gas liberation from the arrester during operation can be hazardous for closed installations.
- The expulsion-type arrester has a limited voltage range (up to 33 kV).
- The V-I characteristic of an expulsion-type arrester is not ideal, so it is not suitable for the protection of expensive devices.

8. **Valve Type Lightning Arresters**

The valve-type arresters are made of series spark gaps and nonlinear resistors. The spark gaps are not conducting under normal voltage, but under surge voltage, they are conducting.

The nonlinear resistor is made of an inorganic compound such as thyrite or metrosil, which under high-frequency surges shows low resistance, but under power frequency, it shows high resistance [11, 13]. Therefore, during surges of overvoltage, the spark gaps are conducted, and the low resistance of the resistor allows the surge current to flow to ground. After the surge is over, the nonlinear resistor offers high resistance to power frequency current to protect the system against a short circuit. Figure 2.9 shows a schematic representation of a valve-type arrester.

Advantages:

- Proper and effective operation against lightning surges.
- Fast response to overvoltage surges.
- The valve-type surge arrester has an impulse ratio almost equal to unity. Therefore, it can withstand overvoltage with a magnitude equal to the rated voltage of the network, making it a reliable protection equipment for networks

Disadvantages/Limitations:

- The valve-type arrester cannot provide effective protection against a very fast-fronted surge wave.
- Moisture may affect the valve-type arrester operation.

9. **Thyrite Lightning Arrester**

This kind of arrester is made of a series of gaps that are in series with thyrite-type resistors. Figure 2.10 shows the thyristor lightning arrester.

The thyrite resistor is made of several discs, and each disc is made of a special inorganic material called thyrite. The thyrite resistor shows high resistance at low

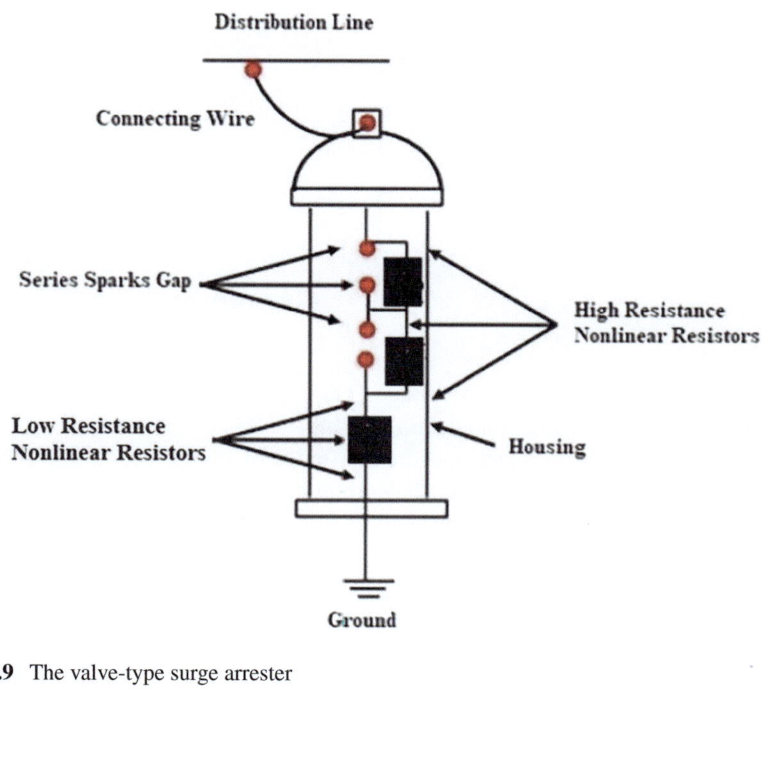

Fig. 2.9 The valve-type surge arrester

Fig. 2.10 The thyrite type arrester

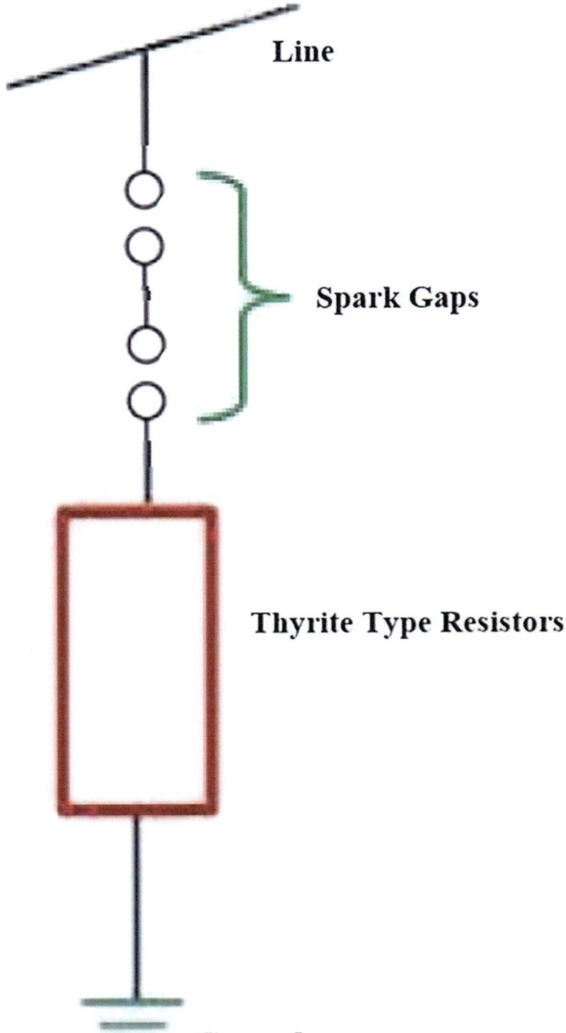

voltage and low resistance at high voltage [11, 13]. These discs are coated with a metal layer and placed inside a porcelain housing.

During surges, the gaps spark and the resistance of thyrite decreases, allowing surge current to pass to ground. After the current passes to ground, the resistance of the thyrite resistor increases and returns to its original state, and the system returns to its regular operation.

10. **Auto Valve Arrester**

The auto valve arrester is made of porous material such as porcelain, which is separated by thin mica rings in the form of flat discs. The discs are not made of uniform composition, and they are made of a mixture of non-conducting and conducting materials [11, 13]. If the surge voltage across the arrester exceeds the threshold value, the discharge starts and the voltage is reduced to a specific value (typically 350 V per disc).

11. **Oxide Film Arrester**

The oxide film arrester is made of certain dry chemicals whose electrical conductivity changes due to the application of heat. For example, lead peroxide has "a specific resistance of 1 Ω per inch cube at ordinary temperatures. When the temperature rises to around 150 °C, the lead peroxide is converted to red lead, which has a specific resistance of around 24 million ohms per inch cube [11].

The oxide film arrester has several cells, each of which has two metal electrodes, and the space between the electrodes is filled with lead peroxide powder. Under surge voltage, the film breaks down at several points, and discharge occurs, leading to the change of lead peroxide to red lead due to heat, preventing the power current from flowing to earth. The typical voltage range of each cell is 250–300 V.

12. **Metal Oxide Lightning Arrester**

One of the most common surge arresters, which is widely used in power system protection against lightning surges, is a metal oxide surge arrester [14].

A metal oxide surge arrester is made of a stack of several ZnO discs. Figure 2.11 shows a schematic representation of this kind of arrester.

The ZnO disc's resistance is very nonlinear, and this nonlinearity is for an extensive current range; therefore, this kind of arrester can be used without a gap [11, 13, 14]. This is the reason that it is also called a gapless arrester.

The voltage across the ZnO arrester at normal working conditions should not exceed the rated voltage of the stack of ZnO discs; therefore, at normal operation conditions, the ZnO arrester shows a high resistance to ground. When a surge comes on the arrester, the resistance decreases quickly and the current flows to the ground, protecting the equipment. As the voltage on the arrester returns to its normal value, the resistance of the arrester increases, and the system returns to regular operation.

i. Advantages:

- The ZnO arrester can pass high current.
- The ZnO arrester has an excellent protection ability.
- The ZnO arrester has a proper sealing condition [15].
- The ZnO arrester is in proper mechanical condition.
- The ZnO arrester is in good decontamination condition.
- The ZnO arrester has a high reliability performance.
- The ZnO arrester has a good power frequency voltage withstand capability [15].

ii. Disadvantages/Limitations:

- The ZnO surge arrester has a specific energy capability and may be damaged during multiple-stroke lightning.

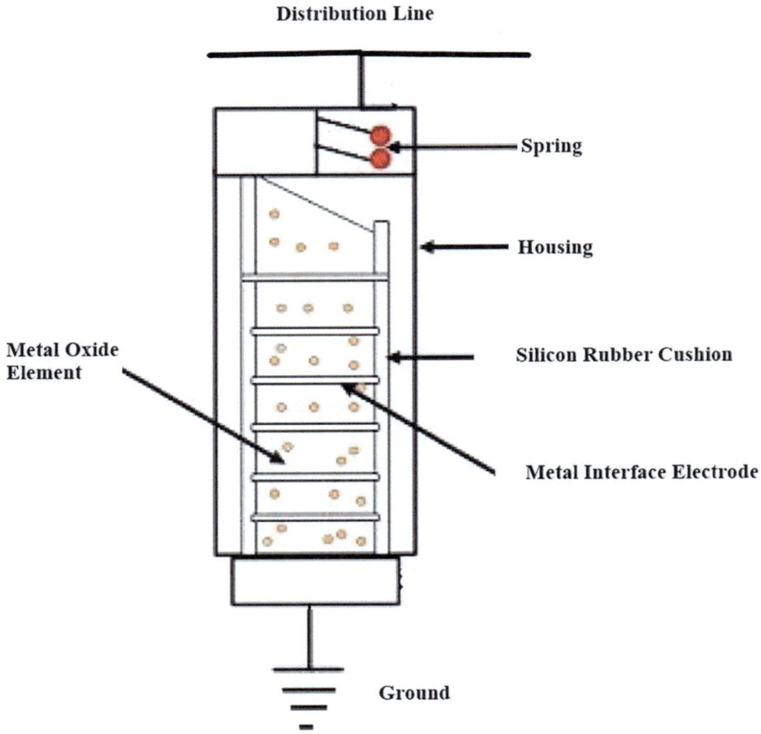

Fig. 2.11 The schematic representation of ZnO arrester

- The ZnO surge arrester for long-time operation may experience aging, and its performance may be reduced.
- The space requirements for installation of ZnO surge arrester may be a limitation for using it in electrical panels or equipment rooms [16]

2.5 How to Protect the Distribution Line Against Lightning

As we know, one of the significant causes of the distribution line interruption is lightning [17]. The severity of the lightning effect on the distribution line is related to the ground flash density of lightning. For higher ground flash density, the probability of a lightning problem for the distribution line is higher. Lightning can interrupt the distribution line in three ways: (i) by direct striking the overhead distribution line, (ii) by striking the shield wire or tower and cause back flashover, (iii) by striking the object near to line and causing induced voltage on distribution line which may lead to flashover (which is the subject of this book).

One of the methods to reduce the effect of lightning on the distribution line is using a surge arrester. The Second method for reducing the impact of lightning on the distribution line is using shield wire. The third method is increasing the insulation level of the line.

One of the primary methods for protecting overhead lines is using a surge arrester. The surge arrester can protect the line from direct and indirect lightning (induced voltage) strokes by limiting the voltage level below the flashover level. Different factors may affect the performance of an arrester in the protection of a distribution line. The essential factor is the spacing of the arrester. Other factors, such as ground resistance, lead length, and arrester energy capability, have less effect on the protection performance of the arrester. In each phase, the surge arrester will be installed, and by installing the surge arresters close to each other, the performance of the line against lightning is better. From the point of view of theory, if surge arresters are installed on each pole, no flashover occurs. On the other hand, if the distance between arresters increases (more than every 5–6 poles), the performance of the line against lightning is decreased. If combination methods are used, the performance of the line against lightning is increased.

Nowadays, different surge arresters are used for the protection of distribution lines, but among them, ZnO surge arresters are widely used in distribution systems [18]; in this case, the reduction of ground resistance is not critical. Because when a surge comes on the distribution line, the arrester will be conductive, and during activation time, the residual voltage of the arrester is on the line, and line protection can occur.

2.5.1 Location of Surge Arresters

Surge arresters are generally installed on the load side of distribution lines. Surge arresters may be installed on every distribution line pole in order to improve the level of protection against lightning [18].

> The choice of lightning protection devices is based on the following parameters and conditions: in addition to considering the structure of the regional network, the characteristics of the equipment and other factors, but also in conjunction with the characteristics of the local lightning catastrophe and the degree of damage to lines and equipment. [18]

2.5.2 Advantages of Lightning Protection of Distribution Lines

The distribution lines are an essential part of the electrical network, and the safe operation of distribution lines is related to the reliable operation of the power system. A well-designed lightning protection system for the distribution system, according to local area conditions, and proper comparison and selection of protective devices can reduce lightning faults in the distribution system, which effectively improves the reliability of the power distribution system. The ZnO surge arresters are economical with relatively stable performance, and they are suitable for distribution lines protection.

References

1. Electrical Terminology Team, What happens if the voltage is too high. https://www.electricalterminology.com/what-happens-if-the-voltage-is-too-high
2. L. Ren, Illuminating the dark side: lightning's assault on power networks. https://clouglobal.com/illuminating-the-dark-side-lightnings-assault-on-power-networks/
3. V. Cooray (ed.), *Lightning Protection* (IET, UK, 2010)
4. P. Chowdhuri, *Electromagnetic Transients in Power Systems*, 2nd edn. (Prentice Hall, India, 2008)
5. L. Ren, Illuminating the dark side: lightning's assault on power networks (2023). https://clouglobal.com/illuminating-the-dark-side-lightnings-assault-on-power-networks/
6. Al. Martin, Lightning damage to equipment without a metallic connection to an external communications service (2011). https://incompliancemag.com/article/lightning-damage-to-equipment-without-a-metallic-connection-to-an-external-communications-service/
7. L. Coghlan, How to avoid power surge damage to your vital systems and components (2021). https://lightningmaster.com/surge-damage/
8. Eaton, Surge protection for electrical power installations (2016). https://www.mtl-inst.com/images/uploads/AN_1009_Rev_B.pdf
9. Renown Electrical System Pvt. Ltd, Types of lightning arrester (2024). https://renownpower.com/types-of-lightning-arrester.Jan
10. Electrical Desk, Types of lightning arresters (2022). https://www.electricaldesks.com/2022/12/types-of-lightning-arresters.html
11. Circuit Globe, Types of lightning arresters (2024). https://circuitglobe.com/types-of-lightning-arresters.html
12. Faadoo Engineers, Multi gap arrester (2023). http://www.faadooengineers.com/online-study/post/eee/switchgear-and-protection/525/multigap-arrester
13. Electrical Work Book, What is a multi-gap arrester? (2021). https://electricalworkbook.com/multi-gap-arrester/
14. V. Hinrichsen, *Metal-Oxide Surge Arresters Fundamentals*, 1st edn. (Siemens AG, Germany, 2001)
15. Huazheng Electric Manufacturing (Baoding) Co., What are the advantages of zinc oxide arrester (2021). https://www.electric-test.com/info/what-are-the-advantages-of-zinc-oxide-arrester-61778914.html
16. RF Wireless World, What is surge arrester (2023). https://www.rfwireless-world.com/Terminology/Surge-arrester-construction-working-types-advantages-disadvantages.html
17. J. McDaniel, Lightning protection of distribution lines: things to consider for a reliable design, in *2005/2006 IEEE/PES Transmission and Distribution Conference and Exhibition* (Dallas, TX, USA, May 2006)
18. ZMS Cable, Protection measures against lightning in aerial cables (2024). https://zmscable.es/en/proteccion-contra-rayo-en-cables-aereos/

Chapter 3
Literature Review on Lightning-Induced Voltage

Abstract Induced voltage on distribution system due to lightning is the main part of lightning damage to distribution system. In computation of induced voltage followings are important:

(i) Choosing a proper model for return stroke channel current.
(ii) Computation of electromagnetic field on the line due to lightning.

In this chapter a literature review on two above mentioned subject is conducted.

3.1 Introduction

Lightning is one of the leading causes of equipment damage in distribution systems. More than 90% of blackouts in overhead distribution systems are due to lightning [1]. Induced voltage on the distribution system due to lightning is the central part of this damage. In the computation of induced voltage, the following are essential:

(i) Choosing a proper model for the return stroke channel current.
(ii) Computation of the electromagnetic field on the line due to lightning.

3.2 Return Stroke Channel

At the time of electrical discharge (when the electric field around the negative cloud is 10 kV/cm, which is equal to the required electric field for breakdown in the air around the cloud), ionized channels start from the cloud to the ground. Still, most of these channels are not continuing anymore due to following two reasons:

(i) The negative charges of the channel are neutralized with positive charges around the channel.
(ii) The cloud charge is not enough to support them in reaching the ground.

Channel continues step by step, and 40 microseconds after the first step, the second one starts, which goes through the same trace as the first one passed, and at each step,

B. Vahidi and A. F. Nematolahi, *Lightning Induced Overvoltage and Current on Distribution Power Systems*, https://doi.org/10.1007/978-981-95-3712-9_3

the channel goes 50 m toward the ground [2]. When the channel reaches enough to the ground the charge from the ground start to move upward and the downward and the upward charge touch each other and a high current move upward to neutralize the channel charges, this current is called return stroke current and sometimes is about 200 kA, but many references indicated that the average value of this current is 20 kA [3].

3.2.1 Models of Return Stroke Channel

Four model types can be considered for the return stroke channel. They are [4]:

(i) The physical or gas-dynamic model, which is analyzed radially, is a small part of the lightning channel.
(ii) The distributed circuit models can be considered approximated electromagnetic circuits. In these models, lightning discharge is regarded as a transient phenomenon on a vertical transmission line with resistance of R, inductance of L, and capacitance of C per unit length.
(iii) The engineering models which describe the space–time distribution of channel current (charge density) according to return stroke channel characteristics such as channel base current, upward velocity of channel, and channel shape.
(iv) The electromagnetic models are usually based on a lossy antenna. These models contain numerical solutions to Maxwell's equations to find the current distribution across the channel. By these models, electric and magnetic fields can be computed far from the channel. These models are more accurate than other models for modeling the lightning channel in the software.

Essentially, the gas-dynamic model is used for reproducing the physical parameters of the return stroke, but other models are used for reproducing the electromagnetic field of the return stroke. Most of the methods that are introduced for computing the induced overvoltage are based on engineering models.

3.2.2 Engineering Models

An engineering model consists of a mathematical characteristic of the location-time of the current i (z', t) or charge density. $\rho(z', t)$. In a lightning channel. These mathematical characteristics include return stroke channel velocity as model input, charge distribution along the channel, and some other adjustable discharge parameters which can be evaluated by comparing with laboratory results. In these models, the return stroke channel is considered vertical to the perfect conducting ground. From an engineering point of view, the best models for return stroke current are those that can be related to the current of the base of the channel ($i(0, t)$), because the current of the base of the channel is measured directly, and related laboratory information about it is available. In the published papers, the mathematical characteristics of the location-time of the current along the channel are introduced as Eq. (3.1) [5]:

$$i(z', t) = u\left(t - \frac{z'}{v_f}\right)P(z')i\left(0, t - \frac{z'}{v}\right) \tag{3.1}$$

where

u step function
v_f wave velocity (return stroke velocity).
v propagation velocity of return stroke current
$P(z')$ damping function of return stroke current along the channel.

Also, for the charge density on the channel, the following continuity equation can be written. The charge density per unit length $\rho_L^*(z', t)$ at height z' and at time t' related to the current $i^*(z', t)$ With the following equation.

$$\frac{\partial \rho_L^*(z', t')}{\partial t'} = -\frac{\partial i^*(z', t')}{\partial z'} \tag{3.2}$$

Star sign (*) means that the variables are selected in a way that they are non-zero everywhere [5].

The custom engineering models for return stroke current are:

(i) Bruce and Golde (BG) model
(ii) Travel current source model (TCS)
(iii) Transmission line model by Uman and Mc Lane
(iv) Modified transmission line model by Rakov.
(v) Modified transmission line model by Nucci and Rachidi.

3.2.3 Current of Return Stroke at the Channel Base

In some countries, the current at the channel base is measured by devices that are located on the transmission line towers and are available [6]. The occurrences and rate of rise of positive lightning are less than negative lightning; therefore, usually negative lightning is considered for strike analysis. For the current of return stroke at the channel base, several models are introduced in the published papers, which are as follows [7].

3.2.3.1 Bruce and Golde Model

This model introduces the current of return stroke at the channel base is a double exponential waveshape. Bruce and Golde introduced Eqs. (3.3) and (3.4) for the current of return stroke at the channel base for the first and subsequent strokes.

Table 3.1 The values of parameters for the current of return stroke at the channel base (Bruce and Golde Model)

Parameter	First stroke	Second stroke
I_0(kA)	30	15
$\alpha[s^{-1}]$	4.4×10^4	4.4×10^4
$\beta[s^{-1}]$	4.6×10^5	4.6×10^5

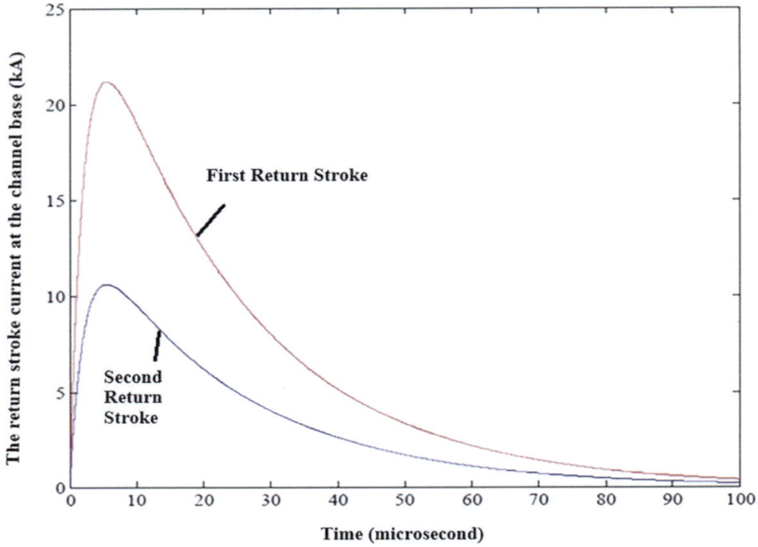

Fig. 3.1 The current of return stroke at the channel base for first and second stroke according to Bruce and Golde model

$$i_{\text{first}}(0, t) = I_0(e^{-\alpha t} - e^{-\beta t}) \tag{3.3}$$

$$i_{\text{second}}(0, t) = \frac{I_0}{2}(e^{-\alpha t} - e^{-\beta t}) \tag{3.4}$$

I_0 = Peak of the current of the return stroke at the channel base

The values of parameters of Eqs. (3.3) and (3.4) are mentioned in Table 3.1.

Figure 3.1 shows the current of the return stroke at the channel base for the first and second strokes according to the Bruce and Golde model.

3.2.3.2 Pierce and Cianos (PC) Model

Pierce and Cianos introduced Eqs. (3.5) and (3.6) for the current of return stroke at the channel base of the first and second stroke [7].

Table 3.2 The values of parameters for the current of return stroke at the channel base (Pierce and Cianos model)

Parameter	First stroke	Second stroke
I_0(kA)	30	15
I_{0i}(kA)	2	2
$\alpha[s^{-1}]$	4.4×10^4	4.4×10^4
$\beta[s^{-1}]$	4.6×10^5	4.6×10^5
$\gamma[s^{-1}]$	10^3	10^3
$\rho[s^{-1}]$	10^3	10^4

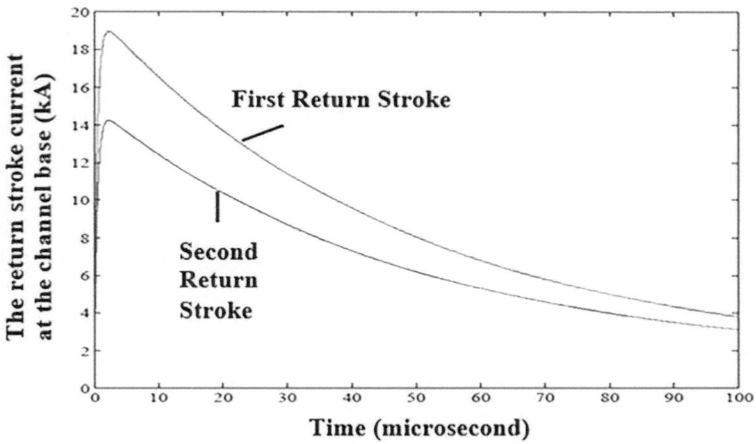

Fig. 3.2 The current of return stroke at the channel base for the first and second strokes according to Pierce and Cianos' model

$$i_{\text{first}}(0, t) = I_0(e^{-\alpha t} - e^{-\beta t}) + I_{0i}(e^{-\gamma t} - e^{-\delta t}) \tag{3.5}$$

$$i_{\text{second}}(0, t) = \frac{I_0}{2}(e^{-\alpha t} - e^{-\beta t}) + \frac{I_{0i}}{2}(e^{-\gamma t} - e^{-\delta t}) \tag{3.6}$$

The values of parameters of Eqs. (3.5) and (3.6) are mentioned in Table 3.2 [7].

Figure 3.2 shows the current of the return stroke at the channel base for the first and second strokes according to Pierce and Cianos' model.

3.2.3.3 Heidler Model

This model for the current of the channel base will produce a return stroke current waveshape with a concave rising portion and will not exhibit a discontinuity in its time derivative, as the double exponential waveshape shows. According to this model, Eq. (3.7) introduced the waveshape for the current of the return stroke at the channel base.

Table 3.3 Suggested parameters by Heidler for channel base current

I_0 (kA)	τ_1 (μs)	τ_2 (μs)	n
10.7	0.25	2.5	2

$$i(0,t) = \frac{I_0}{\eta} \frac{\left({}^{t}/_{\tau_1}\right)^n}{1 + \left({}^{t}/_{\tau_1}\right)^n} \exp\left(-{}^{t}/_{\tau_2}\right) \tag{3.7}$$

$$\eta = \exp\left[\left(-{}^{\tau_1}/_{\tau_2}\right)\left({}^{n\tau_2}/_{\tau_1}\right)^{{}^{1}/_{n}}\right] \tag{3.8}$$

where

τ_1 time constant of front wave
τ_2 time constant of wave tail
I_0 peak of channel base current
n an integer number between 2 and 10.

The following values are suggested for the above parameters by Heidler (Table 3.3) [8].

The Heidler model is the most usable return stroke current model for induced voltage computation.

3.3 Field Computation Around Return Stroke Channel Above Ground

In most models, when the time-place distribution of current along the return stroke channel is known, electromagnetic fields can be computed. In all of the engineering models for the return stroke channel for the computation of induced voltage, the electromagnetic fields should be computed first.

The dipole method is used to compute electromagnetic fields around the return stroke channel. This method computes the electromagnetic fields by assuming that the channel is made of several small current elements, and since continuity exists, each element is considered as a small dipole.

Till now, several different methods have been introduced for the computation of fields of return stroke channel; these methods consist of methods of the frequency domain [9] or the time domain [10–12].

Recently, a method was introduced in reference [13], which accelerates the computation. All of these methods suggest that the ground is lossless and perfectly conductive.

Figure 3.3 shows the geometry of induced voltage computation on a line.

A dipole is shown in Fig. 3.4 to compute the electromagnetic fields. The equations for this dipole near a lossless ground in the time domain will be discussed as

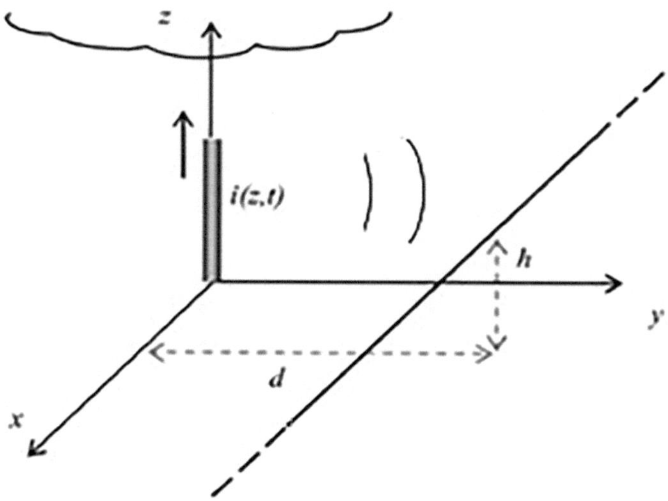

Fig. 3.3 Geometry of induced voltage computation

follows. For computation, the ground is assumed to be perfectly conducting [14]. The calculation is performed in cylindrical coordinates (Fig. 3.4).

$$\mathrm{d}E_p(p,z,t) = \frac{p(z-z\prime)}{4\pi\varepsilon_0}\left[\frac{3}{cr^4}i\left(z\prime,t-\frac{r}{c}\right)+\frac{1}{c^2r^3}\frac{\partial}{\partial t}i\left(z\prime,t-\frac{r}{c}\right)\right.$$
$$\left.+\frac{3}{r^5}\int\limits_0^t i\left(z\prime,t-\frac{r}{c}\right)\mathrm{d}\tau\right] \tag{3.9}$$

$$\mathrm{d}E_z(p,z,t) = \frac{\mathrm{d}z\prime}{4\pi\varepsilon_0}\left[\begin{array}{c}\frac{2(z-z\prime)^2-p^2}{cr^4}i\left(z\prime,t-\frac{r}{c}\right)-\frac{p^2}{c^2r^3}\frac{\partial}{\partial t}i\left(z\prime,t-\frac{r}{c}\right)\\ +\frac{2(z-z')^2-p^2}{r^5}\int\limits_0^t i\left(z',t-\frac{r}{c}\right)\mathrm{d}\tau\end{array}\right] \tag{3.10}$$

$$\mathrm{d}H_\phi = \frac{p\mathrm{d}z\prime}{4\pi r^2}\left[\frac{1}{r}i\left(z\prime,t-\frac{r}{c}\right)+\frac{1}{c}\frac{\partial}{\partial t}i\left(z\prime,t-\frac{r}{c}\right)\right] \tag{3.11}$$

With $R = \sqrt{r^2-(z-z\prime)^2}$ And by using the image method for computing the total field, the above equations should be integrated over $-H$ to H (H is the channel length), then in the time and frequency domains are as follows:

Time domain:

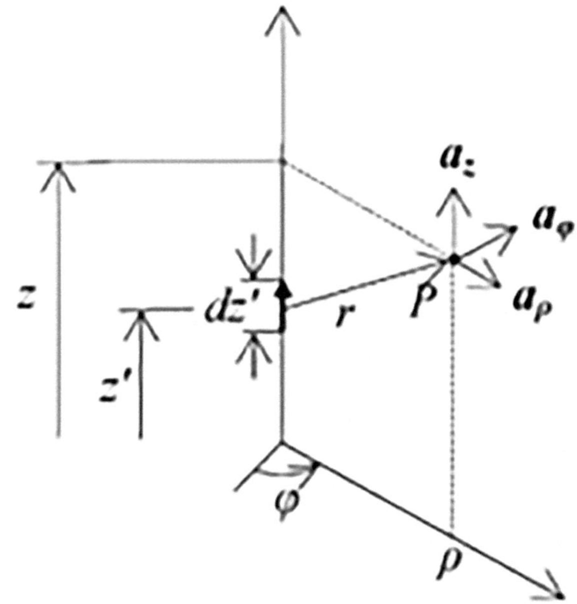

Fig. 3.4 The electromagnetic fields in cylindrical coordinates

$$E_{ziL} = \int_{-H}^{H} \left[\frac{p(z - z\prime)}{4\pi \varepsilon_0} \left[\begin{array}{l} \frac{3}{cr^4} i\left(z\prime, t - \frac{r}{c}\right) \\ + \frac{1}{c^2 r^3} \frac{\partial}{\partial t} i\left(z\prime, t - \frac{r}{c}\right) + \frac{3}{r^5} \int_0^t i\left(z\prime, t - \frac{r}{c}\right) \mathrm{d}\tau \end{array} \right] \mathrm{d}z \right] \tag{3.12}$$

$$E_{riL} = \int_{-H}^{H} \left[\frac{1}{4\pi \varepsilon_0} \left[\begin{array}{l} \frac{2(z-z\prime)^2 - p^2}{cr^4} i\left(z\prime, t - \frac{r}{c}\right) - \frac{p^2}{c^2 r^3} \frac{\partial}{\partial t} i\left(z\prime, t - \frac{r}{c}\right) \\ + \frac{2(z-z\prime)^2 - p^2}{r^5} \int_0^t i\left(z\prime, t - \frac{r}{c}\right) \mathrm{d}\tau \end{array} \right] \mathrm{d}z \right] \tag{3.13}$$

$$H_{\phi iL} = \int_{-H}^{H} \left[\frac{p\prime}{4\pi r^2} \left[\frac{1}{r} i\left(z\prime, t - \frac{r}{c}\right) + \frac{1}{c} \frac{\partial}{\partial t} i\left(z\prime, t - \frac{r}{c}\right) \right] \mathrm{d}z \right] \tag{3.14}$$

Frequency domain:

$$E_{ziL} = \frac{I(0, \omega)}{4\pi \varepsilon_0} \int_{-H}^{H} \left[\frac{2(z - z\prime) - r^2}{CR^4} + \frac{2(z - z\prime) - r^2}{-j\omega R^5} + j\omega \frac{r^2}{c^2 R^3} \right] \mathrm{e}^{j\omega\left(\frac{R}{C} + \frac{|z\prime|}{v}\right)} P(z\prime) \mathrm{d}z\prime \tag{3.15}$$

$$E_{riL} = \frac{I(0,\omega)}{4\pi\varepsilon_0}\int_{-H}^{H}\left[\frac{3(z-z\prime)}{CR^4} + \frac{3r(z-z\prime)}{-j\omega R^5} + j\omega\frac{r(z-z\prime)}{cR^3}\right]e^{j\omega\left(\frac{R}{c}+\frac{|z\prime|}{v}\right)}P(z\prime)\mathrm{d}z\prime \tag{3.16}$$

$$H_{\phi iL} = \frac{I(0,\omega)}{4\pi}\int_{-H}^{H}\left[\frac{r}{R^3} - j\omega\frac{r}{cR^2}\right]e^{j\omega\left(\frac{R}{c}+\frac{|z\prime|}{v}\right)}P(z\prime)dz\prime \tag{3.17}$$

It should be noted that the fields' components in the above equations are composed of three parts: (i) electrostatic fields, (ii) induced fields, and (iii) radiated fields.

3.4 Effect of Lossy Ground

The integral of Maxwell's equations generally has two parts: one part is the direct term, which is for an unlayered environment, and the other part is for a layered environment, which appears as an infinite integral of Bessel functions, which is called the Sommerfeld integrals. Computation of these integrals takes time and is an essential problem in numerical electromagnetics.

Computation of Sommerfeld integrals has started from the beginning of the twentieth century till now. For the first time, Sommerfeld did an analysis of dipole radiation over a lossy ground [15, 16]. The applied studies in the field of lossy ground were continued by others [17–20].

As the numerical methods improved, some researchers tried to use these methods for the solution of Sommerfeld integrals [21–28].

Several research studies have been done to find approximate solutions of these integrals in order to compute the Sommerfeld integrals efficiently [28–31]. Recently, Delfino et al. [32, 33] introduced a full-wave method for the accurate solution of Sommerfeld integrals, which can be used for the computation of electromagnetic fields over a lossy ground. It should be noted that before their method, there was no accurate and effective method for solving Sommerfeld integrals in the computation of radiated fields from a return stroke channel over a lossy ground.

3.4.1 Cooray-Rubinstein Approximation

Cooray [34] introduced a method for the computation of the horizontal component of the electric field on the ground surface with finite conduction, and he used the phrase " surface impedance" in his method. The horizontal component of the electric field on the ground surface can be obtained from the surface impedance phrase with the aid of the rotational components of the magnetic field as follows:

$$E_{\mathrm{r}}(r, 0, j\omega) = B_{\phi}(r, 0, j\omega)\frac{c}{\sqrt{\varepsilon_0\varepsilon_{\mathrm{rg}} + \frac{\sigma_{\mathrm{g}}}{j\omega\varepsilon_0}}} \tag{3.18}$$

where

$B_{\phi}(r, 0, j\omega)$ The Fourier transform of the rotational component of the induced magnetic field on the ground surface.

$E_{\mathrm{r}}(r, 0, j\omega)$ Is the Fourier horizontal component of the electric field on the ground surface?

Cooray showed that the results of this equation for distances less than 200 m are reasonable [34].

Rubinstein [35–38] introduced a method, and he showed that the horizontal field is made of two parts. The first part is the horizontal electric field component over lossless ground, and the second part is the correction factor multiplied by the magnetic field in order to add the effect of lossy ground. According to this method, the electric field at a distance of r and height of z in the frequency domain is:

$$E_{\mathrm{r}}(r, z, j\omega) = E_{\mathrm{rp}}(r, z, j\omega) - H_{\phi p}(r, 0, j\omega)\frac{1 + j}{\sigma_{\mathrm{g}}\delta_{\mathrm{g}}} \tag{3.19}$$

where

δ_{g} Depth of penetration

σ_{g} Soil conductivity

$H_{\phi p}(r, 0, j\omega)$ Fourier of the rotational component of the induced magnetic field on the ground surface (for lossless ground).

$E_{\mathrm{rp}}(r, z, j\omega)$ Fourier horizontal component of the electric field at the height of z over the ground surface (for lossless ground).

The two basic assumptions in the Rubinstein approximation are: (i) $\sigma_g \gg \omega\varepsilon_0\varepsilon_{rg}$ (ω is the frequency of rotation, ε_0 and ε_{rg} are the permittivity of air and the relative permittivity of ground, respectively.) and (ii) the magnetic field on the ground surface is not affected by ground conductivity.

If the first assumption cannot occur, then Eq. (3.19) can be rewritten as:

$$E_r(r, z, j\omega) = E_{rp}(r, z, j\omega) - H_{\phi p}(r, 0, j\omega)\frac{c\mu_0}{\sqrt{\varepsilon_{rg}} + \sigma_g/j\omega\varepsilon_0} \tag{3.20}$$

The second part of Eq. (3.20) is similar to the equation that Cooray introduced for the computation of the electric field on the surface of the ground; therefore, it is named the Cooray-Rubinstein formula.

3.5 Coupling Model for Induced Voltage Computation

After computing the electromagnetic fields in the space surrounding the lightning channel, a proper coupling method must be applied to the overhead lines to determine the induced voltage and current. For this purpose, two primary approaches are used: strategies based on transmission line theory and full-wave methods.

Within the transmission line theory framework, several key field-to-line coupling models are prominent in the literature. These methods [39] include.

1. Chowdhury-Gross Model (is known as the Chowdhury model)
2. The Agrawal-Price-Gurbaxani model (is known as the Agrawal model)
3. The Taylor-Satterwhite-Harrison model,
4. The Rachidi model
5. The Rusck model.

While they are formulated differently, it has been shown that the Agrawal, Taylor, and Rachidi models are three equivalent procedures that provide the same results.

Furthermore, under the common assumption of a vertical lightning channel, the Rusck model also becomes identical to these three models. Its relative simplicity has made it a valuable and frequently used tool for engineering studies on the effects of lightning on power systems [40]. In contrast, other approaches like the Chowdhury-Gross Model differ from this set of equivalent models and have been observed to present a dependence on the lightning channel length (i.e., cloud altitude) that is difficult to justify.

These transmission line coupling models are based on a few key assumptions: i. The line is uniform along its length. ii. The line response is a transverse electromagnetic (TEM) wave. iii. The cross-sectional dimensions of the line are significantly smaller than the shortest wavelength of the incident electromagnetic pulse.

Given these assumptions, applying these methods may present challenges in more complex scenarios involving multi-phase lines, non-uniform lines, and the physical sag of the conductors.

3.6 Coupling the Electromagnetic Fields on Buried Underground Cable

Galvan and Cooray [41], by aid of an approximated equation, introduced the horizontal electric field underground as well as the propagation constant and coupling to underground cable. Petrache et al. [42] investigated the effects of induced current on underground cable. In their method, at first, they computed the horizontal electric field on the surface of the cable without considering its presence. By using the Agrawal model for coupling, they solved the equations in both time and frequency domains by using the finite element method.

3.7 Coupling Theory and Analysis of Induced Current and Voltage on Underground Cable

As can be seen in Fig. 3.5, a cable is buried underground.

The conductivity and relative permittivity of the ground are σ_g and ε_{rg} respectively.

The conductivity and relative permittivity of cable insulation are σ_i and ε_{ri} respectively.

The vertical component of the electric field underground can be neglected [42–44]. The induced voltage and current on the shield of the cable can be computed according to Cooray et al. transmission line theory in the frequency domain [43].

Figure 3.6 shows a differential equivalent coupling circuit for the cable of Fig. 3.5, which is affected by an external field.

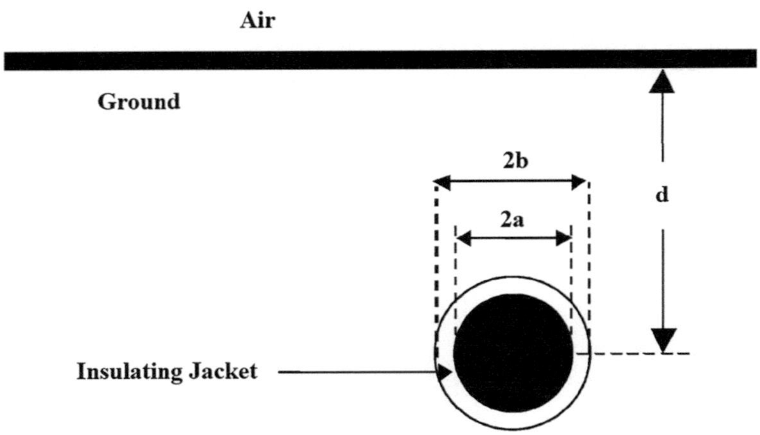

Fig. 3.5 Geometry of the buried cable

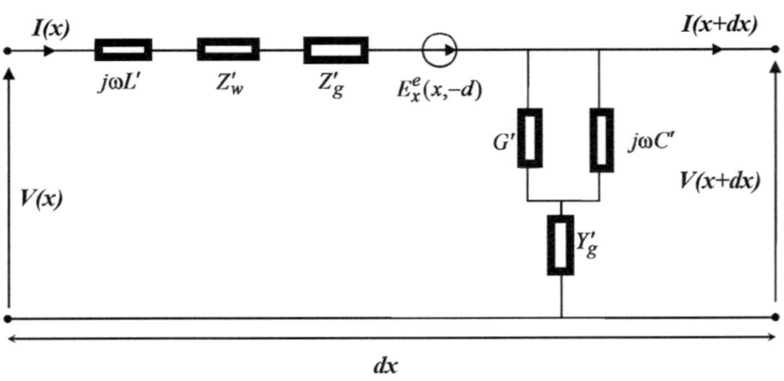

Fig. 3.6 Differential equivalent coupling circuit for an underground cable subjected to an external field

According to Fig. 3.6, we can write the following equations:

$$\frac{\mathrm{d}V(x)}{\mathrm{d}x} + Z'I(x) = E_x^e(x, z = -d) \tag{3.21}$$

$$\frac{\mathrm{d}I(x)}{\mathrm{d}x} + Y'V(x) = 0 \tag{3.22}$$

Longitudinal impedance (Z') and transversal admittance (Y') are as:

$$Z' = j\omega L' + Z'_\omega + Z'_g \tag{3.23}$$

$$Y' = \frac{(G' + j\omega C') \cdot Y'_g}{(G' + j\omega C') + Y'_g} \tag{3.24}$$

L' and C' are the inductance and capacitance of the cable and are computed as follows:

$$L' = \frac{\mu_0}{2\pi} \ln\left(\frac{b}{a}\right) \tag{3.25}$$

$$C' = \frac{2\pi \varepsilon_0 \varepsilon_{rl}}{\ln\left(\frac{b}{a}\right)} \tag{3.26}$$

G' is the conductivity of the cable per unit length and Z'_w Is the internal impedance of the conductor per unit length, Z'_g and Y'_g The impedance and admittance of a unit length of ground. For detailed calculation of the parameters of the circuit of Fig. 3.6, please see the reference [42].

3.8 Numerical Method Solution for Induced Voltage Computation

Before the advent of digital computers, Maxwell's equations were solved by analytical methods, but after 1960, numerical methods were improved in electromagnetic studies with the aid of digital computers. The first series of these methods includes the method of moments (MOM), the finite element method (FEM), and high-frequency methods [45]. Also, these methods have some limitations.

3.8.1 Finite Difference Time Domain

The finite difference time domain (FDTD) method does not use linear algebra or linear equation systems; therefore, it has no limitation on the electrical size of the case. As a source of error in FDTD is well known, they are controllable, and the results of FDTD are accurate and reliable [45]. FDTD solves the problem in the time domain, and for linear systems, the results are like an impulse response, from which the response to any input can be computed in a wide range of frequencies.

As FDTD solves the problem in the time domain, the nonlinear behavior of different elements and devices can be analyzed. The excellent behavior of FDTD as the first method in the time domain was so great that scientists, after the introduction of FDTD, introduced the finite element and finite volume methods in the time domain. Some software that uses FDTD in solving problems is: FEKO (Feldberechnung für Körper mit beliebiger Oberfläche, English translation: field calculations involving bodies of arbitrary shape), CST (Computer Simulation Technology), and HFSS (High-Frequency Structure Simulator).

3.8.2 Different Software

1. CST (Computer Simulation Technology): CST software is generally used for analysis of electromagnetic problems; this software drives the graphical structure based on the Cornell model, automatically makes the mesh structure, and solves the problem. CST allows users to change the mesh structure and simulation method. CST uses different techniques of simulation as follows:

   (i) Solving the transient
   (ii) Solving in the frequency domain
   (iii) Solving the multi-layer problems
   (iv) Solving by the asymptotic method
   (v) Solving by the special mode method.

2. NEC: This software is for simulation and analysis of antennas and uses the MOM for solving the integral equations of the electric field. This software can analyze the electrical field of a thin wire and the magnetic field of conductive surfaces and can be applied to all antenna structures. NEC can analyze the wire structure underground, as well as the insulated conductors and impedance loads.
3. COMSOL: This software simulates problems in the fields of science and engineering, solving equations using the finite element method. This software can add different physics to the model. The user does not need to have high-level knowledge of mathematics or numerical methods. Any model can be simulated by its physical characteristics, such as materials, limitations, or the definition of functions. This software can solve equations in both time and frequency domains.

3.9 Soil Models

3.9.1 Prefect Electric Conductor (PEC)

In this Model, soil is considered lossless, and the computation of fields and induced voltage is based on this assumption. This model is practically the simplest and has fewer complications in computation. Nowadays, this model is used less, but some researchers use this model [46].

3.9.2 Soil with Constant Parameters

In this model, conductivity and other parameters of soil are suggested to be constant. This model is more accurate than the PEC model, but it is still far from the real situation of ground soil.

3.9.3 Multi-layer Soil

In this model, the soil is divided into several layers with different parameters. The number of layers is usually considered to be 2. Layers of soil are considered in horizontal or vertical configurations. This model is more accurate than previous models, but it is still not a perfect model for soil.

3.9.4 Other Recent Works in Soil Modeling

Akbari et al. [47] used the complete wave method based on finite elements to compute the electromagnetic field of the return stroke channel and the induced voltage on the overhead line due to these fields. This paper is one of the first studies to include the frequency dependence of electrical parameters of the soil in induced voltage computation. They used finite elements to evaluate the approximate methods that they used for induced voltage computation. The results of this paper show that only the horizontal component of the electric field at close distances from the return stroke channel is affected by the frequency dependence of the conductivity and permittivity of the soil. For far distances from the return stroke channel, the vertical component of the electric field and the rotational component of the electromagnetic field are also affected by the frequency dependence of the conductivity and permittivity of the soil. The results also show that for soil with high conductivity, the effects of frequency dependence of conductivity and permittivity of soil on induced voltage are less. For soil with low conductivity (less than $\sigma = 0.001$), the effect of frequency dependence of the electrical parameters of soil on induced voltage decreases considerably.

Akbari et al. [48] used the finite element method to analyze the transient behavior of the grounding system, and in their study, they included the frequency-dependent behavior of the electrical parameters of soil. The results show that the frequency dependence of the transient behavior of the grounding system for soil with low conductivity and subjected to subsequent strokes is impressive.

Essentially, the goal of the analysis of the grounding system is due to the fact that induced overvoltage due to lightning is not only affected by coupling between electromagnetic fields and overhead lines, but also affected by conductive currents that transfer to tower grounding through soil (this is called conductive coupling). The results show that when the striking point on the ground is near the line, this conductive coupling has more effects on the induced overvoltage.

Paknahad et al. [49–52] used the full wave method to solve Maxwell's equations using the finite element method to evaluate the electromagnetic fields due to lightning, underground, and overground. They analyzed the induced current on underground cables and the induced voltage on overhead lines. In their study, the effect of layered ground (horizontal and vertical) and the frequency dependence of electrical parameters of soil are included. The results for induced current on buried underground cable are compared with the experimental results of the International Center for Lightning Research and Testing (ICLRT) at Camp Blanding, Florida. The comparison shows that the theoretical results are reasonable. These studies show that the frequency-dependent behavior of the electrical parameters of the soil can increase or reduce the induced current on the shield of an underground cable. The layered ground has an impressive effect on the components of the electromagnetic field underground and affects the induced current on the shield of underground cables. In these studies, the effect of the configuration of ground layers and the vicinity of the power system to the ocean is also analyzed.

Riz et al. [53] investigated the induced current on the shield of the underground cable of the system, and the effect of the conductivity of the soil was studied.

Rizk et al. [54] investigated the effects of a lightning strike on a wind turbine on three-phase lines in the vicinity of the turbine. In this study, the ground is lossy and consists of two horizontal layers, and the effect of the conductivity of each layer and their combination is studied.

Moini et al. [55] investigated the effect of the incline return stroke channel by the method of moments in the presence of lossless ground, and the induced voltage on the overhead line is computed.

References

1. N. Rameli, M.Z.A. AbKadir, M. Izadi, C. Gomes, J. Jasni, Evaluation of lightning induced overvoltage due to variations of channel angle, in *Power Engineering and Optimization Conference (PEOCO), 2013 IEEE 7th International* (IEEE, 2013), pp. 184–187

2. A. Andreotti, D. Assante, F. Mottola, L. Verolino, An exact closed-form solution for lightning-induced overvoltages calculations. IEEE Trans. Power Delivery **24**, 1328–1343 (2009)
3. A. Borghetti, A. Morched, F. Napolitano, C.A. Nucci, M. Paolone, Lightning-induced overvoltages transferred through distribution power transformers. IEEE Trans. Power Delivery **24**, 360–372 (2009)
4. Y. Baba, V.A. Rakov, Electromagnetic models of the lightning return stroke. J. Geophys. Res. Atmos. **112** (2007)
5. M. Paolone, Modeling of lightning-induced voltages on distribution networks for the solution of power quality problems, and relevant implementation in a transient program, Ph.D. thesis, Department of Electrical Engineering, University of Bologna, Bologna, Italy, 2001
6. T. Udo, Estimation of lightning current wave front duration by the lightning performance of Japanese EHV transmission lines. IEEE Trans. Power Delivery **8**, 660–671 (1993)
7. F. Mottola, Methods and techniques for the evaluation of lightning induced overvoltages on power lines. Application to MV distribution systems for improving the quality of power supply, Ph.D. thesis, Department of Electrical Engineering, University Federico II of Napoli, Napoli, Italy, 2008
8. C.A. Nucci, Tensions induites par la foudre sur les lignes aériennes de transport d'énergie; Partie I: Modèles de courant d'arc en retour avec un courant spécifié à la base du canal pour l'évaluation des champs électromagnétiques correspondants, Electra 161, pp 74–102, août 1995
9. S. Ramo, J.R. Whinnery, *Fields and Waves in Modern Radio* (Wiley, NY, USA, 1945)
10. H.J. Price, A.K. Agrawal, The response of a transmission line illuminated by lightning-induced electromagnetic fields. IEEE Trans. Electromagn. Compat. **22**, 150–156 (1980)
11. A.K. Agrawal, H.J. Price, S.H. Gurbaxani, Transient response of multiconductor transmission lines excited by a nonuniform electromagnetic field. IEEE Trans. Electromagn. Compat. **22**, 119–129 (1980)
12. M.A. Uman, D.K. McLain, E.P. Krider, The electromagnetic radiation from a finite antenna. Am. J. Phys. **43**, 33–38 (1975)
13. F. Napolitano, An analytical formulation of the electromagnetic field generated by lightning return strokes. IEEE Trans. Electromagn. Compat. **53**, 108–113 (2011)
14. F. Rachidi, C.A. Nucci, M. Ianoz, C. Mazzetti, Influence of a lossy ground on lightning-induced voltages on overhead lines. IEEE Trans. Electromagn. Compat. **38**, 250–264 (1996)
15. A. Sommerfeld, *Electrodynamics: Lectures on Theoretical Physics* (Academic Press, Cambridge, Massachusetts, USA, 1952)
16. A. Sommerfeld, *Partial Differential Equations in Physics* (Academic Press, Cambridge, Massachusetts, USA, 1949)
17. K.A. Norton, The propagation of radio waves over the surface of the earth and in the upper atmosphere. Proceedings of the Institute of Radio Engineers **25**, 1203–1236 (1937)
18. I.M. Longman, Note on a method for computing infinite integrals of oscillatory functions, in *Mathematical Proceedings of the Cambridge Philosophical Society* (Cambridge University Press, 1956), pp. 764–768
19. A. Baños, *Dipole Radiation in the Presence of a Conducting Halfspace* (Pergamon, Oxford, UK, 1966)
20. J.R. Wait, Electromagnetic surface waves. Adv. Radio Res. **1**, 157–217 (1964)
21. K.A. Michalski, Extrapolation methods for Sommerfeld integral tails. IEEE Trans. Antennas Propag. **46**, 1405–1418 (1998)
22. P. Parhami, Y. Rahmat-Samii, R. Mittra, An efficient approach for evaluating Sommerfeld integrals encountered in the problem of a current element radiating over lossy ground. IEEE Trans. Antennas Propag. **28**, 100–104 (1980)
23. P. Parhami, R. Mittra, Wire antennas over a lossy half-space. IEEE Trans. Antennas Propag. **28**, 397–403 (1980)
24. R. Mittra, P. Parhami, Y. Rahmat-Samii, Solving the current element problem over lossy half-space without Sommerfeld integrals. IEEE Trans. Antennas Propag. **27**, 778–782 (1979)
25. T.J. Cui, W.C. Chew, Modeling of arbitrary wire antennas above ground. IEEE Trans. Geosci. Remote Sens. **38**, 357–365 (2000)

26. J.R. Wait, Propagation of radio waves over a stratified ground. Geophysics **18**, 416–422 (1953)
27. M. Siegel, R.W.P. King, Electromagnetic fields in a dissipative half-space: a numerical approach. J. Appl. Phys. **41**, 2415–2423 (1970)
28. T. Sarkar, Analysis of radiation by arrays of parallel vertical wire antennas over plane imperfect ground (Sommerfeld formulation). IEEE Trans. Antennas Propag. **24**, 544–545 (1976)
29. P.R. Bannister, New formulas that extend Norton's farfield elementary dipole equations to the quasi-nearfield range, Naval Underwater Systems Center New London CT (1984)
30. C.E. Baumann, E.E.S. Sampaio, Electric field of a horizontal antenna above a homogeneous half-space: implications for GPR. Geophysics **65**, 823–835 (2000)
31. K.A. Michalski, J.R. Mosig, Multilayered media Green's functions in integral equation formulations. IEEE Trans. Antennas Propag. **45**, 508–519 (1997)
32. F. Delfino, R. Procopio, M. Rossi, Lightning return stroke current radiation in presence of a conducting ground: 1. Theory and numerical evaluation of the electromagnetic fields. J. Geophys. Res. Atmos. **113** (2008)
33. F. Delfino, R. Procopio, M. Rossi, F. Rachidi, C.A. Nucci, Lightning return stroke current radiation in presence of a conducting ground: 2. Validity assessment of simplified approaches. J. Geophys. Res. Atmos. **113** (2008)
34. V. Cooray, Horizontal fields generated by return strokes. Radio Sci. **27**, 529–537 (1992)
35. M. Rubinstein, M.A. Uman, P.J. Medelius, E.M. Thomson, Measurements of the voltage induced on an overhead power line 20 m from triggered lightning. IEEE Trans. Electromagn. Compat. **36**, 134–140 (1994)
36. F. Rachidi, M. Rubinstein, S. Guerrieri, C.A. Nucci, Voltages induced on overhead lines by dart leaders and subsequent return strokes in natural and rocket-triggered lightning. IEEE Trans. Electromagn. Compat. **39**, 160–166 (1997)
37. M. Rubinstein, An approximate formula for the calculation of the horizontal electric field from lightning at close, intermediate, and long range. IEEE Trans. Electromagn. Compat. **38**, 531–535 (1996)
38. M. Rubinstein, A.Y. Tzeng, M.A. Uman, P.J. Medelius, E.M. Thomson, An experimental test of a theory of lightning-induced voltages on an overhead wire. IEEE Trans. Electromagn. Compat. **31**, 376–383 (1989)
39. F. Rachidi, Formulation of the field-to-transmission line coupling equations in terms of magnetic excitation field. IEEE Trans. Electromagn. Compat. **35**, 404–407 (1993)
40. V. Cooray, G. Cooray, M. Rubinstein, F. Rachidi, Exact expressions for lightning electromagnetic fields: application to the Rusck field-to-transmission line coupling model. Atmosphere **14**, 350 (2023)
41. A. Galvan, V. Cooray, Lightning induced voltages on bare and insulated buried cables, in *13th International Zurich Symposium on Electromagnetic Compatibility*, Zurich, Switzerland (1999)
42. E. Petrache, F. Rachidi, M. Paolone, C.A. Nucci, V.A. Rakov, M.A. Uman, Lightning induced disturbances in buried cables-part I: theory. IEEE Trans. Electromagn. Compat. **47**, 498–508 (2005)
43. V. Cooray, Underground electromagnetic fields generated by the return strokes of lightning flashes. IEEE Trans. Electromagn. Compat. **43**, 75–84 (2001)
44. E.F. Vance, *Coupling to Shielded Cables* (Wiley Press, NJ, USA, 1978)
45. A. Taflove, K.R. Umashankar, The finite-difference time-domain method for numerical modeling of electromagnetic wave interactions. Electromagnetics **10**, 105–126 (1990)
46. F. Napolitano, A. Borghetti, C.A. Nucci, F. Rachidi, M. Paolone, Use of the full-wave finite element method for the numerical electromagnetic analysis of LEMP and its coupling to overhead lines. Electric Power Systems Research **94**, 24–29 (2013)
47. M. Akbari, K. Sheshyekani, A. Pirayesh, F. Rachidi, M. Paolone, A. Borghetti, C.A. Nucci, Evaluation of lightning electromagnetic fields and their induced voltages on overhead lines considering the frequency dependence of soil electrical parameters. IEEE Trans. Electromagn. Compat. **55**, 1210–1219 (2013)

48. M. Akbari, K. Sheshyekani, M.R. Alemi, The effect of frequency dependence of soil electrical parameters on the lightning performance of grounding systems. IEEE Trans. Electromagn. Compat. **55**, 739–746 (2013)
49. J. Paknahad, K. Sheshyekani, M. Hamzeh, D. Li, F. Rachidi, The influence of the slope angle of the ocean–land mixed propagation path on the lightning electromagnetic fields. IEEE Trans. Electromagn. Compat. **57**, 1086–1095 (2015)
50. J. Paknahad, K. Sheshyekani, F. Rachidi, M. Paolone, A. Mimouni, Evaluation of lightning-induced currents on cables buried in a lossy dispersive ground. IEEE Trans. Electromagn. Compat. **56**, 1522–1529 (2014)
51. J. Paknahad, K. Sheshyekani, F. Rachidi, M. Paolone, Lightning electromagnetic fields and their induced currents on buried cables, Part II: the effect of a horizontally stratified ground. IEEE Trans. Electromagn. Compat. **56**, 1146–1154 (2014)
52. J. Paknahad, K. Sheshyekani, F. Rachidi, Lightning electromagnetic fields and their induced currents on buried cables, Part I: the effect of an ocean–land mixed propagation path. IEEE Trans. Electromagn. Compat. **56**, 1137–1145 (2014)
53. M.E.M. Rizk, F. Mahmood, M. Lehtonen, E.A. Badran, M.H. Abdel-Rahman, Investigation of lightning electromagnetic fields on underground cables in wind farms. IEEE Trans. Electromagn. Compat. **58**, 143–152 (2016)
54. M.E.M. Rizk, F. Mahmood, M. Lehtonen, E.A. Badran, M.H. Abdel-Rahman, Induced voltages on overhead line by return strokes to grounded wind tower considering horizontally stratified ground. IEEE Trans. Electromagn. Compat. **58**, 1728–1738 (2016)
55. R. Moini, S.H.H. Sadeghi, B. Kordi, F. Rachidi, An antenna-theory approach for modeling inclined lightning return stroke channels. Electr. Power Syst. Res. **76**, 945–952 (2006)

Chapter 4
Identifying Lightning Channel-Base Current Characteristics and Finding Heidler Function Parameters

Abstract As Heidler's model is a good approximation for current at channel base this model is used widely in lightning studies and be adopted by IEC 62305. The Heidler's model is used by many researchers. Heidler's function has its own parameters and there is not a specific method for adjustment of these parameters for a known values lightning current parameters. Heidler et-al. introduced a graphical method with graphs which parameters of Heidler's model can be determined from them, but using these graphs when the numbers of currents increase is challenging. In this chapter a Cholesky-based Monte Carlo is used to generate the characteristics of lightning current in order to statistically analysis the lightning.

4.1 Introduction

In recent decades, the use of electronic devices has increased rapidly, and as they are sensitive to lightning transients, researchers are focused on the electromagnetic interference of the lightning electromagnetic field with these devices. This interference damages the quality and reliability of electrical and electronic devices. This interference is analyzed by simulation and researchers do their best those conditions and simulation parameters are close enough to reality [1–24].

Lightning is a random phenomenon and using a statistical model for analyzing the lightning is necessary [25]. In the analysis of distribution system, the lightning outage of system dependents on lightning overvoltage which exceeds the CFO of the line [26–30]. The first step in the analysis of the lightning effect is choosing a proper model for return stroke channel current wherein one of the useful model is the engineering model [31, 32]. In engineering model, the spatial time distribution of the channel current is based on the velocity of the channel, the channel current at the base [33]. In the engineering model, the channel base current will reduce as a function of height and time, therefore choosing a proper model and parameters for channel base current are close to reality that can increase the accuracy of computation; more details about engineering models can be found in [34, 35].

B. Vahidi and A. F. Nematolahi, *Lightning Induced Overvoltage and Current on Distribution Power Systems*, https://doi.org/10.1007/978-981-95-3712-9_4

The channel base current and velocity of return stroke can be measured; therefore, all models are improved based on these two parameters. The measured channel base current can be for first stroke or subsquent strokes [35]. The current waveshape has five important parameters:

(i) I_p = Current peak
(ii) $S_m = \mathrm{d}i/\mathrm{d}t$ (maximum rate of rise of current
(iii) t_f or t_{cr} = Front time or time to crest
(iv) t_h = Time to half peak or tail time
(v) Q = Channel charges.

The details of these parameters can be found in [33]. From measurements results around the world which have been done on characteristics of current at channel base can judge that there is a strong relationship between front time, peak of current, rate of rise of current, and tail time of current at channel base. Generally, the parameters of current have a log normal distribution [36]. Several formulas are introduced for current at the base of stroke channel; these formulas should describe the characteristics and waveshape of the current at the base of the channel [37–39]. Different models are introduced for current at the channel base and Heidler's model is the best [40].

As Heidler's model is a good approximation for current at the channel base this model is used widely in lightning studies and be adopted by IEC 62305 [41].

The Heidler's model is used by [19, 42, 43]. Heidler's function has its parameters, and there is no specific method for the adjustment of these parameters for known values of I_p, T_m, S_m, and T_h. Heidler et al. [40] introduced a graphical method with graphs whose parameters of Heidler's model can be determined from them. However, using these graphs when the numbers of currents increase is challenging.

For the Heidler function, the S_m ($\mathrm{d}i/\mathrm{d}t$) is correlated to I_p (current peak) and t_f front time; therefore, this makes adjusting the function parameters more challenging. Due to this many researchers tried to find the parameters in a way that the current waveshape at the base of the channel be the same as they want [42, 44, 45], in [42] and [44] the PSO is used and in [45] the Genetic Algorithm (GA) is used to find the parameters of Heidler's function.

Due to the correlation between parameters, their distribution function is a log-normal function with several variables [26, 46]. In this chapter, a Cholesky-based Monte Carlo is used to generate the characteristics of lightning current in order to statistically analyze the lightning that is extracted from a Ph.D. thesis [47].

4.2 Distribution of Parameters of Multi-variable Normal Function and Cholesky-Based Monte Carlo Method

Many researchers in different articles measuring lightning current showed that the probability density distribution function of the parameters is log-normal, and these parameters are highly correlated. Therefore, finding data for analyzing the multi-variable density distribution function by Monte-Carlo is necessary. The multi-variable density distribution function of current at the channel base is as Eq. (4.1) [47, 48].

$$f(x_1, \ldots, x_n) = \frac{1}{\sqrt{(2\pi)^n |k|}} \exp\left(-\frac{1}{2}(x-\mu)^T k^{-1} (x-\mu)\right) \tag{4.1}$$

where x is the matrix of variables (I_p, t_f, t_h, S_m), k is the covariance matrix of the variables, and μ is the natural logarithm of the mean values matrix of variables. The diagonal elements of the covariance matrix k_{ii} are equal to σ_i^2, and the non-diagonal elements are equal to $k_{ij} = p_{ij} \times \sigma_i \times \sigma_j$. The σ_i is the natural logarithm of the standard deviation of the ith variable, and p_{ij} is the correlation between the ith and jth variable, and $|k|$ is also equal to the determinant of the matrix k [48].

In this chapter, the Cholesky method is used for Monte Carlo data generation for the parameters of lightning current. For modeling the multiple correlated variables problems by the Monte Carlo method, the Cholesky decomposition is highly recommended [49–51].

Cholesky decomposition can generate and incorporate those normal parameters that are uncorrelated into normally correlated variables. For developing the correlated variables, the Cholesky decomposition of the covariance matrix should be found, and then it should be multiplied by non-correlated random variables. Cholesky decomposes a Hermitian positive-definite matrix A by using the (4.2) equation.

$$k = L \times L^{\mathrm{T}} \tag{4.2}$$

where;

- L = A lower triangular matrix with positive and fundamental diagonal elements
- L^{T} = The L's conjugate transpose.

By decomposing the covariance k, the lower-triangular L will be obtained. There is a unique Cholesky decomposition for each Hermitian real-valued symmetric positive-definite matrix.

Table 4.1 shows the statistical characteristics of negative current measured at channel base [47], and Table 4.2 shows the correlation (p) between current parameters [47], which are from measurements around the world [52, 53].

Table 4.3 shows the Cholesky decomposition of the covariance of lightning characteristics (matrix L in Eq. (4.2)).

Table 4.1 The statistical characteristics of the negative current measured at the channel base

	I_p	t_f	S_m	t_h
Mean (μ)	30,000	5.60E−06	1.2E+10	7.50E−05
Standard deviation (Log 10) (σ)	0.26	0.31	0.26	0.26

Table 4.2 The correlation (p) between current parameters

	I_p	t_f	S_m	t_h
I_p	1	0.37	0.36	0.56
t_f	0.37	1	– 0.21	0.33
S_m	0.36	– 0.21	1	0.1

Table 4.3 The Cholesky decomposition of the covariance of lightning characteristics

0.26	0	0	0
0.1147	0.288	0	0
0.0936	– 0.09605	0.22274	0
0.1456	0.034367	– 0.01602	0.21204

For generating statistical data of lightning parameters by the Cholesky-based Monte Carlo method, Eq. (4.3) is used.

$$[X]_{4\times n} = \exp([L]_{4\times 4} \times [\text{Rand}]_{4\times n} + \ln([\mu]_{4\times n})) \tag{4.3}$$

where:

- Rand is a matrix of natural random numbers with the range of [0, 1].
- In Eq. (4.3), $[X]$ is the characteristic that is randomly generated using correlated data.
- $[L]$ is the Cholesky decomposition of the covariance matrix (Table 4.3).
- n is the generated scenarios for Monte Carlo.
- $[\mu]$ is the mean value matrix (each column is the mean of each parameter).

After generating n scenarios for Monte Carlo, the parameters for current at the base of the channel are determined according to Heidler's function (Eq. (4.4)).

$$I = \frac{I_1}{\eta_1}\frac{(t/\tau_{11})^{n_1}}{1+(t/\tau_{11})^{n_1}}\exp\left(\frac{-t}{\tau_{12}}\right) \tag{4.4}$$

where

$$\eta_1 = \exp\left[\left(\frac{-\tau_{11}}{\tau_{12}}\right)\left(\frac{n_1\times\tau_{12}}{\tau_{11}}\right)^{1/(n_1)}\right]$$

where

- I_1 is the amplitude of the channel-base current
- τ_{11} is the front-time constant
- τ_{12} is the decay-time constant
- η_1 is the amplitude correction factor
- n_1 is an exponent.

In order to obtain the parameters for the Heidler model, the genetic algorithm code of MATLAB software is used. The objective function is:

$$\mathrm{Obj} = C_1 \times \frac{\left|I_{\mathrm{ph}} - I_{\mathrm{pc}}\right|}{I_{\mathrm{pc}}} + C_2 \times \frac{|T_{\mathrm{fh}} - T_{\mathrm{fc}}|}{T_{\mathrm{fc}}} + C_3 \times \frac{|T_{\mathrm{hh}} - T_{\mathrm{hc}}|}{T_{\mathrm{hc}}} \tag{4.5}$$

While;

$$\frac{\left|I_{\mathrm{ph}} - I_{\mathrm{pc}}\right|}{I_{\mathrm{pc}}} < 0.005, \frac{|T_{\mathrm{fh}} - T_{\mathrm{fc}}|}{T_{\mathrm{fc}}} < 0.005, \frac{|T_{\mathrm{hh}} - T_{\mathrm{hc}}|}{T_{\mathrm{hc}}} < 0.01, \frac{|S_{\mathrm{mh}} - S_{\mathrm{mc}}|}{S_{\mathrm{mc}}} < 0.2 \tag{4.6}$$

where;

- I_{ph} = Peak of current by Heidler model
- I_{pc} = Peak of current by Cholesky-based Monte Carlo. Method
- T_{fh} = Front time of current by Heidler model
- T_{fc} = Front time of current by Cholesky-based Monte Carlo. Method
- T_{hh} = Tail time of the current by the Heidler model
- T_{hc} = Tail time of current by Cholesky-based Monte Carlo. Method
- S_{mh} = Rate of rise of current by Heidler model
- S_{mc} = Rate of rise of current by Cholesky-based Monte Carlo Method

This objective function is for determining the Heidler function parameters in order to minimize the difference between the characteristics of the waveform obtained by the Heidler and the characteristics of the waveform obtained from the Cholesky-based Monte Carlo method

It should be noted that by considering that simply minimizing the objective function, the best optimized parameters are not yielded, but the parameters are acceptable if constraints in Eq. (4.6) are satisfied.

4.3 Generating the Characteristics of Lightning Current by Cholesky-Based Monte Carlo Method

2000 characteristics are generated by the Cholesky-based Monte Carlo method. The standard deviation and mean value of these characteristics are shown in Table 4.4.

Table 4.4 The standard deviation and mean value of lightning current

	I_{p}	t_{f}	t_{h}	S_{m}
Mean	30,452	5.713E−06	7.610E−05	1.236E+10
Standard deviation	0.259	0.307	0.259	0.263

The correlation between lightning current characteristics generated by the Cholesky-based Monte Carlo method is shown in Table 4.5.

Tables 4.4 and 4.5 are much closer to Tables 4.1 and 4.2, respectively.

Figure 4.1 shows the histogram for the distribution of lightning current characteristics, which is generated by the Cholesky-based Monte Carlo method.

Figure 4.2 shows the congestion of the generated lightning current characteristics by the Cholesky-based Monte Carlo in order to show the correlation between them.

Table 4.5 The correlation between lightning current characteristics generated by the Cholesky-based Monte Carlo method

	I_p	t_f	S_m	t_h
I_p	1.000	0.354	0.356	0.553
t_f	0.354	1.000	– 0.209	0.324
S_m	0.356	– 0.209	1.000	0.092
t_h	0.553	0.324	0.092	1.000

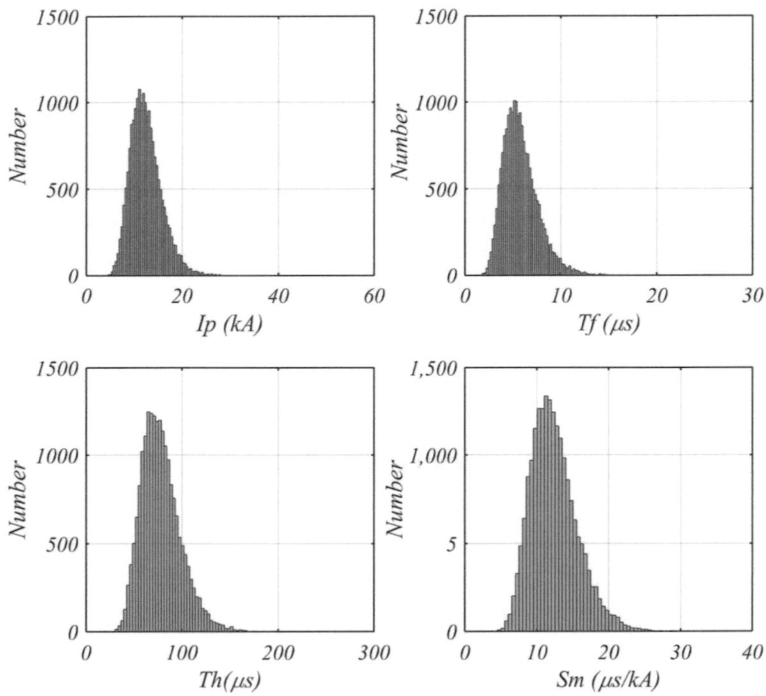

Fig. 4.1 The histogram for the lightning current characteristics distribution, which is generated by the Cholesky-based Monte Carlo method

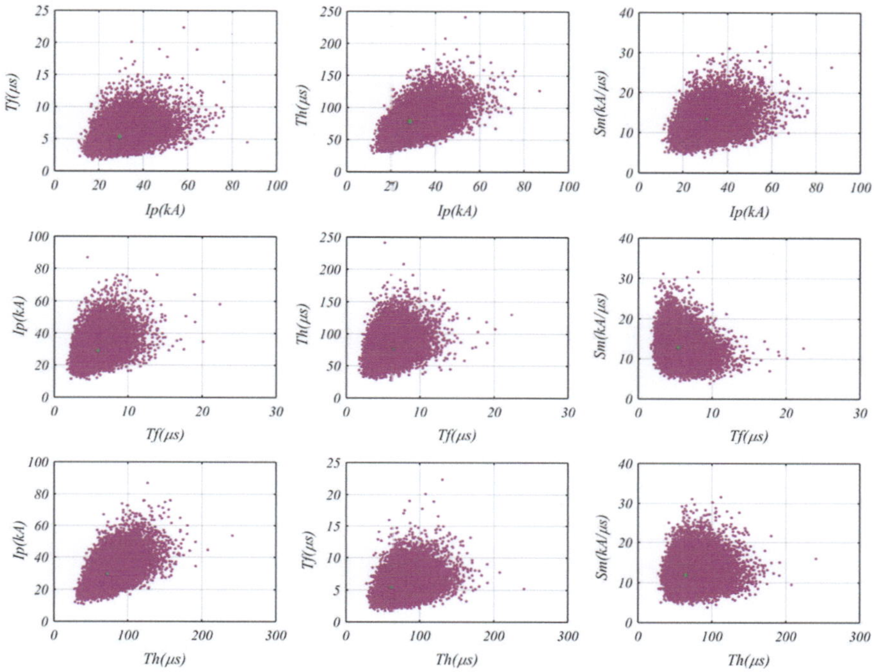

Fig. 4.2 The congestion of the generated lightning current characteristics by the Cholesky-based Monte Carlo

4.4 Extracting Heidler Parameters Related to Each Characteristic of Lightning Current

After the generation of lightning characteristics using the Cholesky method, we can compute the equivalent parameters of the Heidler model for each lightning current using the Genetic Algorithm method. The population and search agents in the Genetic Algorithm method are 100 and 20. The C_1, C_2, and C_3 coefficients are selected as 50, 50, and 100. After each run of the Genetic Algorithm, the constraints of Eq. (4.6) are controlled. If each of these conditions is not met, the run of the Genetic Algorithm is repeated until all conditions are satisfied. The static value for each condition and objective function of results that the Genetic Algorithm obtains are shown in Table 4.6 [47]. The mean errors I_P, T_f, T_h, and S_m are 6.37E−04, 9.16E−04, 1.31E−05, and 1.39E−01.

Figure 4.3 shows the histogram for the parameters of the proposed Heidler's function. Figure 4.4 shows the congestion of these generated lightning current characteristics.

The statistical characteristics of the Heidler parameters are shown in Table 4.7. Table 4.8 shows the correlation between Heider's function parameters [47].

Table 4.9 shows the Cholesky matrix of the resulting Heidler parameter.

Table 4.6 Analyzing the static parameters of the objective function and each condition of the objective function

	$\frac{\lvert I_{ph}-I_{pc}\rvert}{I_{pc}}$	$\frac{\lvert I_{fh}-I_{fc}\rvert}{I_{fc}}$	$\frac{\lvert I_{hh}-I_{hc}\rvert}{I_{hc}}$	$\frac{\lvert S_{mh}-S_{mc}\rvert}{S_{mc}}$	Obj
Mean	6.37E−04	9.16E−04	1.31E−05	1.39E−01	1.2105
Std	2.78E−03	1.56E−02	5.48E−03	2.62E−01	0.55712

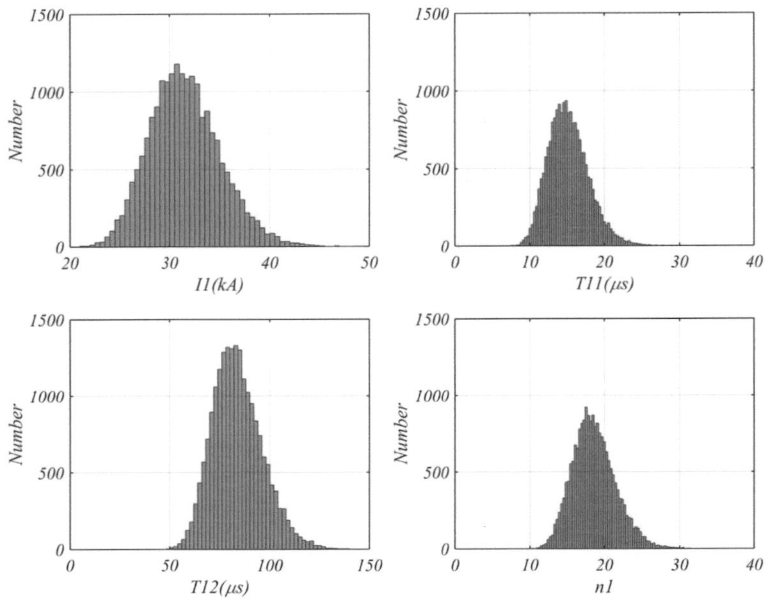

Fig. 4.3 The histogram for the Heidler function is based on Cholesky decomposition

4.5 Comparison of Obtained Heidler Function Parameters Based on the Cholesky-Based Monte Carlo Method with Measurement Data

By using Eqs. (4.3) and (4.4), 20,000 lightning characteristics are generated, and the mean value and standard deviation of lightning parameters are shown in Table 4.10. The correlation between these parameters is shown in Table 4.11 [47].

By comparing the computed parameters and their correlation (Tables 4.10 and 4.11) with measured parameters and their correlation (Tables 4.1 and 4.2), we can see that the measured and computed I_p, t_f, and t_h are close to each other.

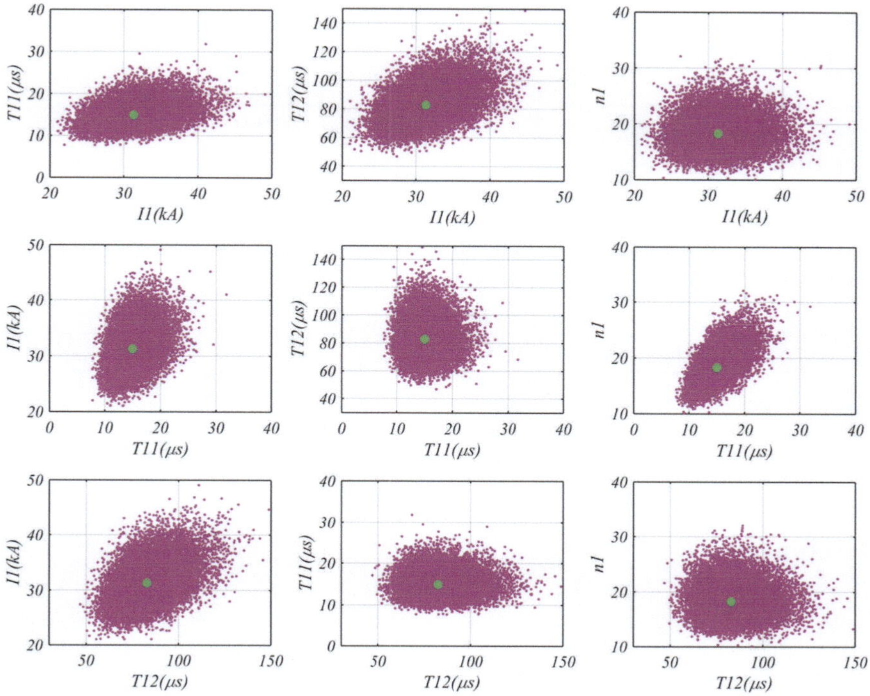

Fig. 4.4 The congestion of the Heidler function parameters based on the Cholesky-based Monte Carlo (to show the correlation between them)

Table 4.7 The statistical characteristics of the Heidler parameters

	I_1 (A)	T_{11} (μs)	T_{12} (μs)	n_1
Mean	31,321.17	14.9415	82.83984	18.32746
Std (log 10)	0.257165	0.406122	0.336767	0.353221

Table 4.8 The correlation between Heilder function parameters

	I_1	T_{11}	T_{12}	n_1
I_1	1	0.339202	0.452634	0.062668
T_{11}	0.339202	1	− 0.00908	0.617352
T_{12}	0.452634	− 0.00908	1	− 0.03877
n_1	0.062668	0.617352	− 0.03877	1

Table 4.9 The Cholesky matrix of the resulting Heidler parameter

0.257165	0	0	0
0.139756	0.381318	0	0
0.148887	− 0.06259	0.295511	0
0.023031	0.231147	0.019776	0.265358

Table 4.10 Statistical specification of lightning characteristics generated by the Cholesky-based Monte Carlo method

	I_p	t_f	t_h	S_m
Mean	30,584.52	5.68E−06	7.84E−05	1.05E+10
Standard deviation (log 10)	0.25801	0.309015	0.269534	0.320939

Table 4.11 Correlation between the lightning characteristics generated by the Cholesky-based Monte Carlo method

	I_p	t_f	S_m	t_h
I_p	1.00	0.38	0.42	0.56
t_f	0.38	1.00	− 0.41	0.32
S_m	0.42	− 0.41	1.00	0.17
t_h	0.56	0.32	0.17	1.00

References

1. M. Akbari, K. Sheshyekani, A. Pirayesh, F. Rachidi, M. Paolone, A. Borghetti, C.A. Nucci, Evaluation of lightning electromagnetic fields and their induced voltages on overhead lines considering the frequency dependence of soil electrical parameters. IEEE Trans. Electromagn. Compat. **55**, 1210–1219 (2013)
2. M. Akbari, K. Sheshyekani, M.R. Alemi, The effect of frequency dependence of soil electrical parameters on the lightning performance of grounding systems. IEEE Trans. Electromagn. Compat. **55**, 739–746 (2013)
3. J. Paknahad, K. Sheshyekani, M. Hamzeh, D. Li, F. Rachidi, The influence of the slope angle of the ocean–land mixed propagation path on the lightning electromagnetic fields. IEEE Trans. Electromagn. Compat. **57**, 1086–1095 (2015)
4. J. Paknahad, K. Sheshyekani, F. Rachidi, M. Paolone, A. Mimouni, Evaluation of lightning-induced currents on cables buried in a lossy dispersive ground. IEEE Trans. Electromagn. Compat. **56**, 1522–1529 (2014)
5. J. Paknahad, K. Sheshyekani, F. Rachidi, M. Paolone, Lightning electromagnetic fields and their induced currents on buried cables, Part II: The effect of a horizontally stratified ground. IEEE Trans. Electromagn. Compat. **56**, 1146–1154 (2014)
6. J. Paknahad, K. Sheshyekani, F. Rachidi, Lightning electromagnetic fields and their induced currents on buried cables, Part I: The effect of an ocean–land mixed propagation path. IEEE Trans. Electromagn. Compat. **56**, 1137–1145 (2014)
7. M.E.M. Rizk, F. Mahmood, M. Lehtonen, E.A. Badran, M.H. Abdel-Rahman, Investigation of lightning electromagnetic fields on underground cables in wind farms. IEEE Trans. Electromagn. Compat. **58**, 143–152 (2016)
8. M.E.M. Rizk, F. Mahmood, M. Lehtonen, E.A. Badran, M.H. Abdel-Rahman, Induced voltages on overhead line by return strokes to grounded wind tower considering horizontally stratified Ground. IEEE Trans. Electromagn. Compat. **58**, 1728–1738 (2016)
9. R. Moini, S.H.H. Sadeghi, B. Kordi, F. Rachidi, An antenna-theory approach for modeling inclined lightning return stroke channels. Electr. Power Syst. Res. **76**, 945–952 (2006)
10. F. Napolitano, A. Borghetti, C.A. Nucci, F. Rachidi, M. Paolone, Use of the full-wave finite element method for the numerical electromagnetic analysis of LEMP and its coupling to overhead line. Electr. Power Syst. Res. **94**, 24–29 (2013)

11. K. Sheshyekani, M. Akbari, Evaluation of lightning-induced voltages on multiconductor overhead lines located above a lossy dispersive ground. IEEE Trans. Power Delivery **29**, 683–690 (2014)
12. K. Sheshyekani, J. Paknahad, Lightning electromagnetic fields and their induced voltages on overhead lines: the effect of a horizontally stratified ground. IEEE Trans. Power Delivery **30**, 290–298 (2015)
13. R. Khosravi, S.H.H. Sadeghi, R. Moini, Electromagnetic field due to lightning strike to a tall tower sitting on a mountainous terrain. IEEE Trans. Electromagn. Compat. **58** (2016)
14. M. Azadifar, P. Dehkhoda, S.H.H. Sadeghi, R. Moini, A Hybrid FDFD–MoM technique for susceptibility evaluation of a transmission line inside a perforated enclosure. IEEE Trans. Electromagn. Compat. **56**, 1474–1479 (2014)
15. J. Zhang, Q. Zhang, W. Hou, L. Zhang, F. Zhou, Y. Ma, Y. Ma, Evaluation of the lightning-induced voltages of multiconductor lines for striking cone-shaped mountain. IEEE Trans. Electromagn. Compat. **61**, 1534–1542 (2019)
16. W. Li, T. Xie, W. Zhang, Z. Zhang, D. Mei, Calculation method and influencing factors of lightning-induced voltage on overhead distribution line, in *IEEE 3rd Information Technology and Mechatronics Engineering Conference (ITOEC)*, pp. 420–425 (2017)
17. H. Chen, Y. Du, M. Yuan, Q.H. Liu, Lightning-induced voltages on a distribution line with surge arresters using a hybrid FDTD–SPICE method. IEEE Trans. Power Delivery **33**, 2354–2363 (2018)
18. Q. Yang, J. Tao, Y. He, B. Zhang, Analysis and suppression measures of lightning transient overvoltage in the signal cable of wind turbines. Wind Energy **21**, 211–225 (2018)
19. F. Rachidi, W. Janischewskyj, A.M. Hussein, C.A. Nucci, S. Guerrieri, B. Kordi, J.-S. Chang, Current and electromagnetic field associated with lightning-return strokes to tall towers. IEEE Trans. Electromagn. Compat. **43**, 356–367 (2001)
20. A.F. Nematollahi, B. Vahidi, Investigating the effect of lightning channel inclination on the induced voltage in overhead lines in the presence of horizontally stratified ground. Electr. Power Syst. Res. **208**, 1–10 (2022)
21. M. Brignone, R. Procopio, D. Mestriner, M. Rossi, F. Delfino, F. Rachidi, M. Rubinstein, Analytical expressions for lightning electromagnetic fields with arbitrary channel-base current—Part I: Theory. IEEE Trans. Electromagn. Compat. **63**, 525–533 (2020)
22. F. Delfino, R. Procopio, M. Rossi, F. Rachidi, Influence of frequency-dependent soil electrical parameters on the evaluation of lightning electromagnetic fields in air and underground. J. Geophys. Res. Atmos. **114**, 1–12 (2009)
23. G.V. Cooray, *The Lightning Flash* (IET, UK, 2003)
24. M.R.B. Tavakoli, B. Vahidi, Transmission-lines shielding failure-rate calculation by means of 3-D leader progression models. IEEE Trans. Power Delivery **26**, 507–516 (2010)
25. A. Rahiminejad, B. Vahidi, J. He, A fractal-based stepped downward leader model including branched channel charge distribution and branch fading. Electr. Power Syst. Res. **176**, 1–10 (2019)
26. A. Borghetti, F. Napolitano, C.A. Nucci, M. Paolone, Effects of nearby buildings on lightning induced voltages on overhead power distribution lines. Electr. Power Syst. Res. **94**, 38–45 (2013)
27. A. Borghetti, C.A. Nucci, M. Paolone, Indirect-lightning performance of overhead distribution networks with complex topology. IEEE Trans. Power Delivery **24**, 2206–2213 (2009)
28. A. Borghetti, C.A. Nucci, M. Paolone, An improved procedure for the assessment of overhead line indirect lightning performance and its comparison with the IEEE Std. 1410 method. IEEE Trans. Power Delivery **22**, 684–692 (2006)
29. A. Borghetti, C.A. Nucci, M. Paolone, Estimation of the statistical distributions of lightning current parameters at ground level from the data recorded by instrumented towers. IEEE Trans. Power Delivery **19**, 1400–1409 (2004)
30. A. Borghetti, F. Napolitano, C.A. Nucci, F. Tossani, Influence of the return stroke current waveform on the lightning performance of distribution lines. IEEE Trans. Power Delivery **32**, 1800–1808 (2016)

31. V.A. Rakov, M.A. Uman, Review and evaluation of lightning return stroke models including some aspects of their application. IEEE Trans. Electromagn. Compat. **40**, 403–426 (1998)
32. C.A. Nucci, G. Diendorfer, M.A. Uman, F. Rachidi, M. Ianoz, C. Mazzetti, Lightning return stroke current models with specified channel-base current: a review and comparison. J. Geophys. Res. Atmos. **95**, 20395–20408 (1990)
33. W.R. Gamerota, J.O. Elismé, M.A. Uman, V.A. Rakov, Current waveforms for lightning simulation. IEEE Trans. Electromagn. Compat. **54**, 880–888 (2011)
34. V. Cooray, Unification of engineering return stroke models. Electr. Power Syst. Res. **195**, 107118 (2021)
35. V. Cooray, M. Rubinstein, F. Rachidi, Modified transmission line model with a current attenuation function derived from the lightning radiation field—MTLD model. Atmosphere **12**, 1–24 (2021)
36. V.A. Rakov, Lightning parameters for engineering applications (keynote speech), in *2010 Asia-Pacific International Symposium on Electromagnetic Compatibility* (IEEE, 2010), pp. 1120–1123
37. V. Javor, P.D. Rancic, A channel-base current function for lightning return-stroke modeling. IEEE Trans. Electromagn. Compat. **53**, 245–249 (2010)
38. D.S. Prasad, G.S. Punekar, Concerning channel base current functions for lightning studies. IEEE Trans. Electromagn. Compat. **54**, 1318–1320 (2012)
39. G. Diendorfer, M.A. Uman, An improved return stroke model with specified channel-base current. J. Geophys. Res. Atmos. **95**, 13621–13644 (1990)
40. F. Heidler, J.M. Cvetic, B.V. Stanic, Calculation of lightning current parameters. IEEE Trans. Power Delivery **14**, 399–404 (1999)
41. F. Heidler, W. Zischank, Z. Flisowski, C. Bouquegneau, C. Mazzetti, Parameters of lightning current given in IEC 62305-background, experience and outlook, in *29th International Conference on Lightning Protection*, p. 26 (2008)
42. G. Yang, F. Zhou, Y. Ma, Z. Yu, Y. Zhang, J. He, Identifying lightning channel-base current function parameters by powell particle swarm optimization method. IEEE Trans. Electromagn. Compat. **60**, 182–187 (2017)
43. A. Andreotti, C. Petrarca, A. Pierno, On the effects of channel tortuosity in lightning-induced voltages assessment. IEEE Trans. Electromagn. Compat. **57**, 1096–1102 (2015)
44. G. Ramarao, K. Chandrasekaran, Evaluating lightning channel-base-current function parameters for identifying interdependence of wavefront and tail by PSO method. IEEE Trans. Electromagn. Compat. **61**, 183–190 (2018)
45. K. Chandrasekaran, G.S. Punekar, Use of genetic algorithm to determine lightning channel-base current-function parameters. IEEE Trans. Electromagn. Compat. **56**, 235–238 (2013)
46. A. Borghetti, A. Morched, F. Napolitano, C.A. Nucci, M. Paolone, Lightning-induced overvoltages transferred through distribution power transformers. IEEE Trans. Power Delivery **24**, 360–372 (2008)
47. A.F. Nematollahi, Invesigation of induced overvoltage caused by lightning in distribution networks considering environmental and soil conditions. Ph.D. thesis, Amirkabir University of Technology, Tehran, Iran (2023)
48. A.F. Nematollahi, B. Vahidi, The application of the Cholesky-based Monte Carlo method to evaluate lightning channel base current. IEEE Trans. Electromagn. Compat. **65**, 804–811 (2023)
49. N. Burgess, Correlated Monte Carlo Simulation using Cholesky Decomposition (2022)
50. M. Asai, M. McAleer, J. Yu, Multivariate stochastic volatility: a review. Economet. Rev. **25**, 145–175 (2006)
51. M. Pourahmadi, M.J. Daniels, T. Park, Simultaneous modelling of the Cholesky decomposition of several covariance matrices. J. Multivar. Anal. **98**, 568–587 (2007)
52. A. Andreotti, A. Pierno, V.A. Rakov, An analytical approach to calculation of lightning induced voltages on overhead lines in case of lossy ground—Part I: Model development. IEEE Trans. Power Delivery **28**, 1213–1223 (2013)
53. IEC Protection against lightning Part 1: General Principles (2011)

Chapter 5
Analyzing the Effect of Incline Return Stroke Channel on the Induced Voltage on the Overhead Line

Abstract Many articles in recent decades investigated the lightning overvoltage. The researches in this field can be categorized according to methods of solving the equations, ground parameters and lightning channel model. In this chapter for analyzing the lightning induced voltage by 3D FEM method, the problem is modeled in RF module of COMSOL Multiphysics software and the schematic method of modeling is shown in Fig. 5.1. In this modeling height of return stroke channel is 2 km and two directions is chosen for slope of channel (θ and φ). θ is the angle of slope with respect to Z axis and φ is the angle of slope with respect to X axis in the XY coordinate plane. The θ angle is varying from 0° to 30° and the φ angle is varying from 0° to 180°. Ground is two vertical layers which consisted of ocean and land.

5.1 Introduction

In recent decades, electronic devices have been used in many applications, and lightning transients can cause malfunctions of these devices. Therefore, in the design step of distribution networks, the effect of lightning transients should be considered. In distribution lines, the induced lightning overvoltage due to indirect stroke is more than the lightning overvoltage due to direct stroke.

5.2 Previous Works

Many articles in recent decades have investigated the lightning overvoltage [1–7]. Research in this field can be categorized according to the methods used to solve the equations, ground parameters, and the lightning channel model. The method of equation solution category contains two sub-categories: (i) analytical methods [8] and (ii) numerical methods such as finite element method (FEM) [9–13], Finite difference time domain (FDTD) [14–17], method of moment (MOM) [18]. For the ground parameters method, different methods and different models are introduced,

B. Vahidi and A. F. Nematolahi, *Lightning Induced Overvoltage and Current on Distribution Power Systems*, https://doi.org/10.1007/978-981-95-3712-9_5

such as: perfect electric conductor (PEC), single layer homogenous lossy soil, vertical or horizontal layer soil [12, 13, 19, 20], and frequency dependent soil. Most of the previous researches are focused on the vertical return stroke channel for simplicity of analysis, but the actual return stroke channel is similar to an inclined channel. Therefore, in recent years, some articles have used an inclined return stroke channel for the computation of induced voltage by analytical or numerical methods [21–23]. In Refs. [21, 24], MOM is used to find the induced voltage on a single conductor overhead line due to an inclined return stroke channel by developing the antenna theory in the presence of a PEC ground. Their results show that channel slope can change the induced voltage amplitude in the presence of PEC ground. In reference [8], an analytical formula was developed for the computation of induced voltage in the presence of an inclined return stroke channel with an arbitrary slope in the presence of a PEC ground. Natsui et al. [23, 25] developed a FDTD three-dimensional model for an inclined return stroke channel to compute induced voltage over PEC ground and a single-layer lossy ground. Their results show that the slope of the return stroke channel has a considerable effect on the induced voltage in comparison to the vertical return stroke channel. In the last decades, protection of distribution lines from the impact of lightning has been an essential challenge for distribution line designers. High current discharges at short times, complex physical behavior, and different environmental conditions make the problem more complex. Therefore, analyzing different situations can be helpful for engineers. In this book, a full wave based on the 3D FEM method is used to solve Maxwell's equations, and the induced voltage due to an inclined return stroke channel on different propagation trace combinations is computed [26, 27].

5.3 Modeling

In this book, for analyzing the lightning-induced voltage by the 3D FEM method, the problem is modeled in the RF module of COMSOL Multiphysics software, and the schematic method of modeling is shown in Fig. 5.1 [26].

In this modeling, the height of the return stroke channel is 2 km, and two directions are chosen for the slope of the channel (θ and φ). θ is the angle of slope with respect to the Z axis, and φ is the angle of slope with respect to the X axis in the XY coordinate plane. The θ angle varies from 0° to 30°, and the φ angle varies from 0° to 180°. The ground consists of two vertical layers, which consist of ocean and land.

In COMSOL, Natural Neumann conditions are applied in the ground surface and at the air interfaces. Natural Dirichlet conditions are used for external boundaries [26, 28, 29]. The air is suggested to be lossless, and a cylinder model is used for modeling the environment that surrounds the channel [26, 29]. As we know, the COMSOL software is based on the FEM method, and knowing the boundary conditions and positions is necessary for solving the problem equations. If the position of external boundaries is not selected too far, they will reflect the waves. If the position of external boundaries is selected too far, the meshes and the computational time of the

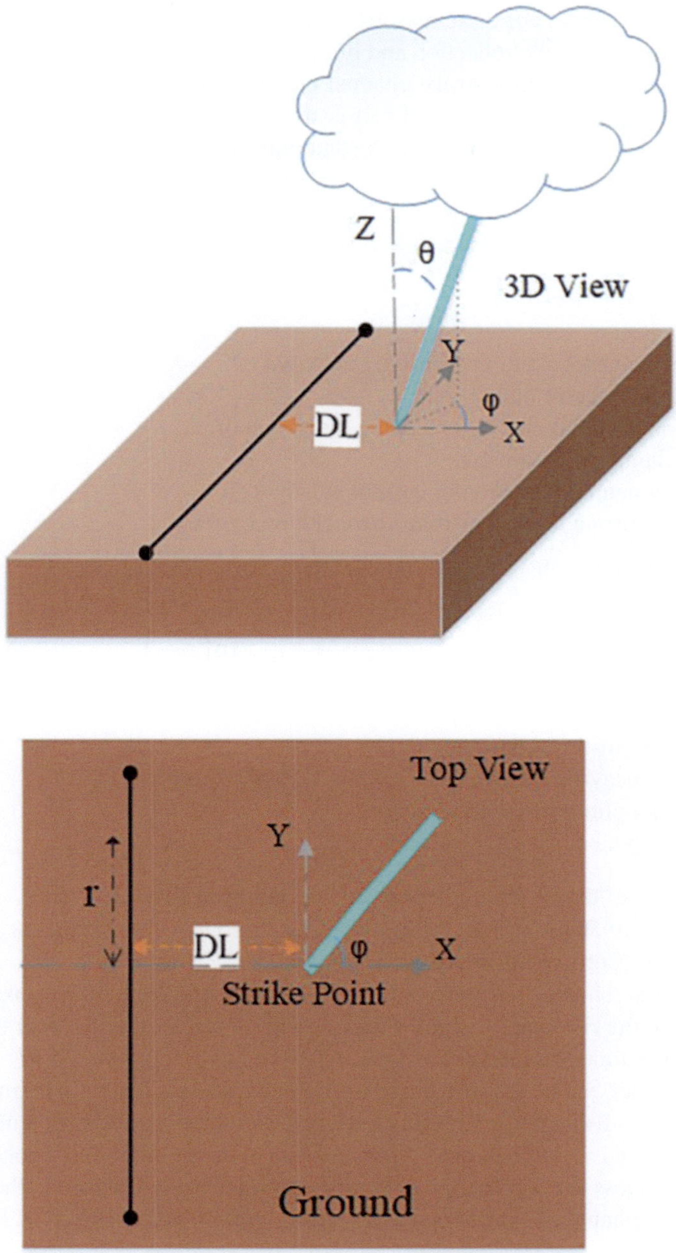

Fig. 5.1 Schematic representation of modelling

problem will increase. For these reasons, scattering boundary conditions are used in COMSOL to avoid wave reflection and limit computation time [26, 28, 29].

For modeling the return stroke channel current, the modified transmission line model with exponential decay (MTLE) is used in this chapter. The return stroke velocity is selected as 1.5e8 m/s, and the damping coefficient is selected as 2000 m. The return stroke current is:

$$I(z,t) = I_0\left(0, t - \frac{z}{v}\right)e^{\frac{-z}{\lambda}} \tag{5.1}$$

where

$I(z, t)$ is the channel current at time t at height z.
v is return stroke velocity.
$\left(t - \frac{z}{v}\right)$ is the delay time for reaching the height of z.
λ is the damping coefficient.
$\left(e^{\frac{-z}{\lambda}}\right)$ exponential reduction of current at height z.
$I_0(0, t)$ is current at return stroke channel base ($z = 0$).

For current at the channel base, the Heidler function is used (Eq. 5.2).

$$I(0,t) = \frac{I_0}{\eta_1}\frac{(t/\tau_{11})^{n_1}}{1 + (t/\tau_{11})^{n_1}}\exp\left(\frac{-t}{\tau_{12}}\right) \tag{5.2}$$

τ_{11} is the front-time constant
τ_{12} is the decay-time constant
η_1 is the amplitude correction factor
n_1 is an exponent.

Parameters of the Heidler function are selected from IEC 62305 [30].

Parameters of 50 kA current for the first and subsequent stroke currents are shown in Table 5.1 (according to IEC 62305).

Current waveshapes are shown in Fig. 5.2. The amplitude of return stroke currents is 50 kA, but the rise times and fall times are chosen differently in order to find the effect of these times on induced voltage.

In this model, the height of the line is 10 m and the line length is 6 km. In Refs. [31, 32], strike distance for lightning current of 50 kA by using different methods is computed, and it is shown that if the distance between the return stroke channel and the line is less than 100 m, the lightning touches the line. Therefore, the distance between the channel and the line is chosen to be 100 m (DL of Fig. 5.1). Figure 5.3

Table 5.1 Parameters for current at base of return stroke channel

Current wave form	I_0 (kA)	τ_1 (μs)	τ_2 (μs)	n_1	η_1
0.25/100	50	0.454	143	10	0.993
1/200	50	1.82	285	10	0.986

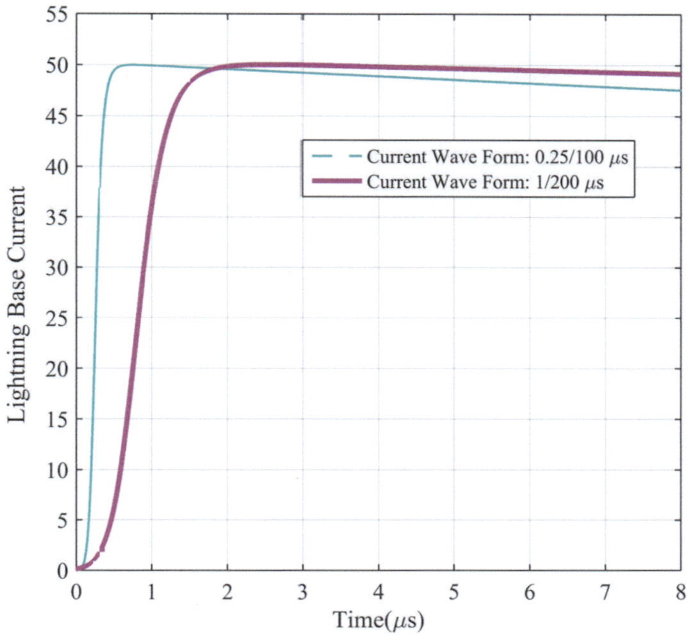

Fig. 5.2 The waveshape of the current (kA) of the channel base

shows the induced voltage due to a vertical return stroke channel over the ground for $\sigma = 0.001$ S/m and $\sigma = 0.01$ S/m.

DL = 100 m and current is 0.25/100 µs. The results of 3D FEM are compared with 3D FDTD method of Ref. [25] at the middle of the line ($r = 0$).

As can be seen, the results of these two methods are matched.

5.4 The Effect of Propagation Over Ground-Ocean on the Induced Voltage on the Overhead Line

5.4.1 The Effect of Vertical Return Stroke Channel on Induced Voltage

Figure 5.4 shows the geometrical presentation of the lightning striking the ground.

In this computation, DL = 100 m, and the overhead line is installed between the return stroke channel and the coastline.

Computation is carried out for two current waveshapes: 0.25/100 µs, 1/200 µs, with a current amplitude of 50 kA. ($x = r =$ distance alongside the line from the middle of the line).

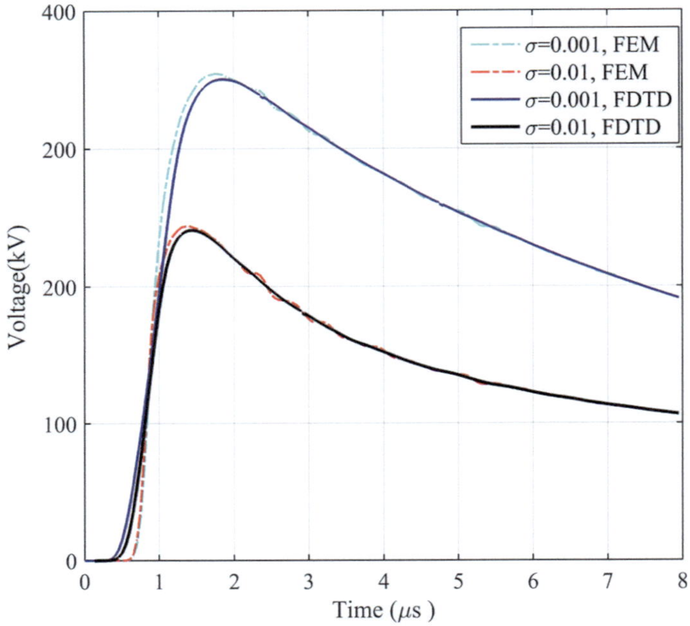

Fig. 5.3 Comparison between the induced voltages of this book method (FEM) and Ref. [25] method (FDTD)

Figure 5.5 shows the induced voltage at: $r = 0$ m, $r = 500$ m and $r = 1000$ m for vertical channel ($\theta = 0°$, $\varphi = 0°$) for DS = 10 m, Ds = 50 m and DS = ∞ (ground is homogenous single layer and $\sigma = 0.001$S/m).

As can be seen at the middle of the line ($r = 0$), the maximum of voltage amplitude occurs at DS = ∞. From Fig. 5.5, it can be seen that for $r = 0$, as DS increases, the induced voltage increases, and the voltage waveshape is closer to the waveshape of the induced voltage over ground. For a faster current waveshape, the induced voltage amplitude is higher due to higher di/dt. From Fig. 5.5 for $r = 500$ m and $r =$ 1000 m, as DS increases the induced voltage amplitude decreases, as reference [33] mentioned the peak and rising time of vertical components of electric fields beneath the line will increase when it is closed to the coastline, and this causes an increase in the voltage peak and rising time.

5.4.2 *The Effect of Inclined Return Stroke Channel on Induced Voltage*

Figure 5.6 shows the induced voltage at the middle of the overhead line for a current waveshape of 0.25/100 μs, $I = 50$ kA.

The peaks of induced voltages are shown in Fig. 5.7.

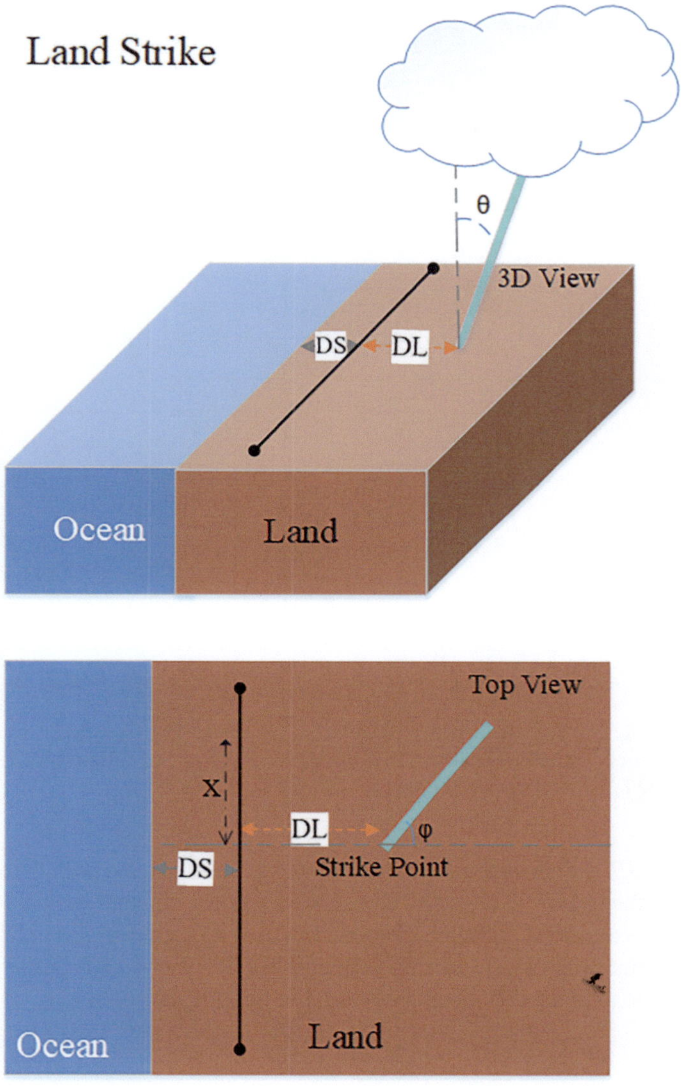

Fig. 5.4 The geometrical presentation of the striking point for induced voltage computation on a single wire overhead line due to an inclined return stroke channel

From Fig. 5.7, it can be judged that the peak of induced voltage for the overhead line that has a distance (DS) of 50 m from the ocean-land interface is higher than that of the same line whose DS is 10 m. From Fig. 5.6, we can judge that by increasing θ, for higher φ, the falling time of the voltage waveshape decreases, and by increasing θ, the falling time decreases.

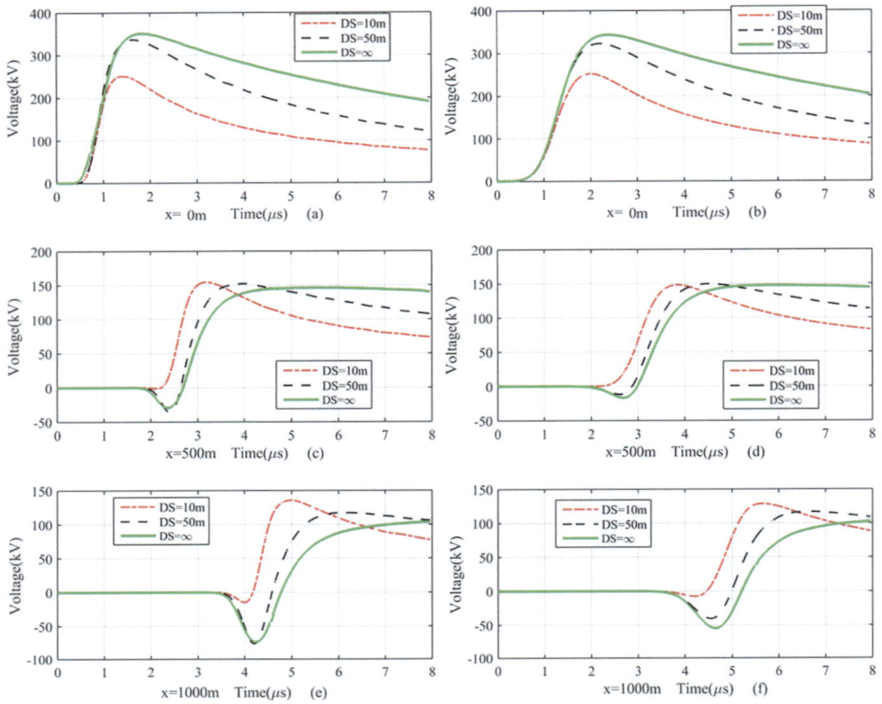

Fig. 5.5 The effect of distance between coastline and overhead line on induced voltage due to vertical return stroke channel (for figures **a**, **c**, and **e**, the waveshape of current is 0.25/100 μs, for figures **b**, **d**, and **f**, the waveshape of current is 1/200 μs, I = 50 kA)

From Fig. 5.7, we can judge that for equal θ, as φ increases, the peak of the induced voltage is higher for $0° < \varphi < 90°$, and at higher θ, the channel induced a voltage with a lower peak in the middle of the line. For $\varphi = 90°$, the variation of θ has little influence on the peak of the induced voltages. For $90° < \varphi < 180°$ by increasing θ, the peak of the induced voltage at the middle of the line increases.

5.4.3 *Induced Voltage Amplitude Along the Overhead Line Due to an Inclined Channel*

Figure 5.8 shows the peak of the induced voltage along the line. As the distance of the observation point from the middle point of the line increases, the peak of the induced voltage decreases. As can be seen from this figure, while the peak of induced voltage at the middle of the line for DS = 50 m is higher than for DS = 10 m, for points farther than the middle point, the peak of voltage for DS = 10 m is higher than for DS = 50 m. This is due to the fact that electromagnetic fields pass over a lossy ground layer, resulting in attenuation of the field.

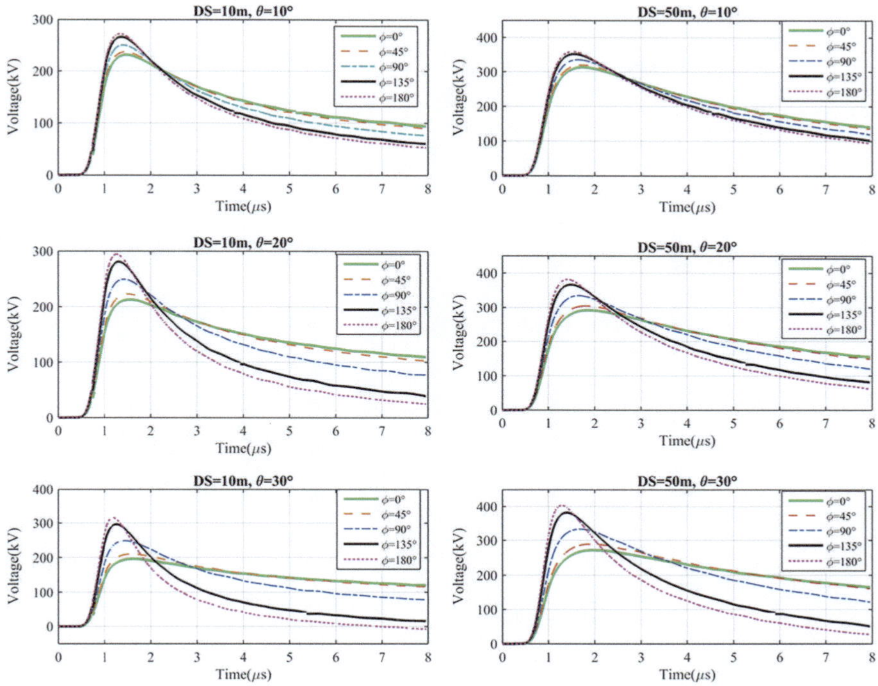

Fig. 5.6 The effect of the angle of slope of the inclined return stroke channel on the induced voltage at the middle of a single overhead line

By comparing the results, we can judge that as the distance between the observation point on the line and the lightning channel increases, the peak of the induced voltage at a distance far from the middle of the line ($r = 1000$ m) reduces. For $\varphi = 90°$, by increasing θ, the distance between the lightning channel and half of the line increases, but the distance between the lightning channel and the other half of the line decreases, resulting in the peak of the induced voltage for the nearer half line being higher than the other half line.

5.5 The Effect of a Strike to the Ocean Surface on Induced Voltage on the Overhead Line

Figure 5.9 shows the geometric configuration. Distance between the middle of the line and the striking point is 100 m (DL = 100 m for Land Side (L.S) or Ocean Side (O.S)). The overhead line height is 10 m above ground and is parallel to the coastline. The simulation is done for DS = 10 m and DS = 50 m.

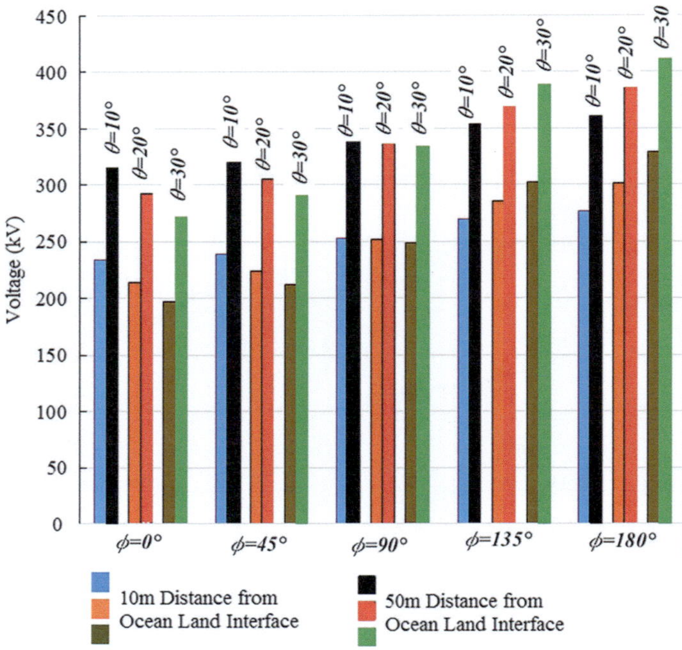

Fig. 5.7 The peaks of induced voltages on a single overhead line due to an inclined return stroke channel

5.5.1 The Effect of the Strike of the Vertical Stroke Channel on the Ocean Surface on the Induced Voltage on the Overhead Line

The induced voltage on the overhead line due to a vertical lightning channel is obtained for a 50 kA lightning current, and two current waveshapes (0.25/100 μs, 1/200 μs). Figure 5.10 shows results for this simulation due to lightning strikes in the ground or ocean [26].

From Fig. 5.10, it can be judged that by increasing the distance between the overhead line and the coastline for both strikes to ground (Land Side, L.S.) or ocean (Ocean Side, O.S), the peak of induced voltage increases.

The waveshapes of induced voltages due to the stroke to ground and ocean for DS = 10 m are very close to each other.

For DS = 50 m stroke to ocean produces induced voltage with a lower peak at the middle of the line with respect to stroke to ground.

From Fig. 5.10, we can judge that for a faster current waveshape, the induced voltage peak is higher.

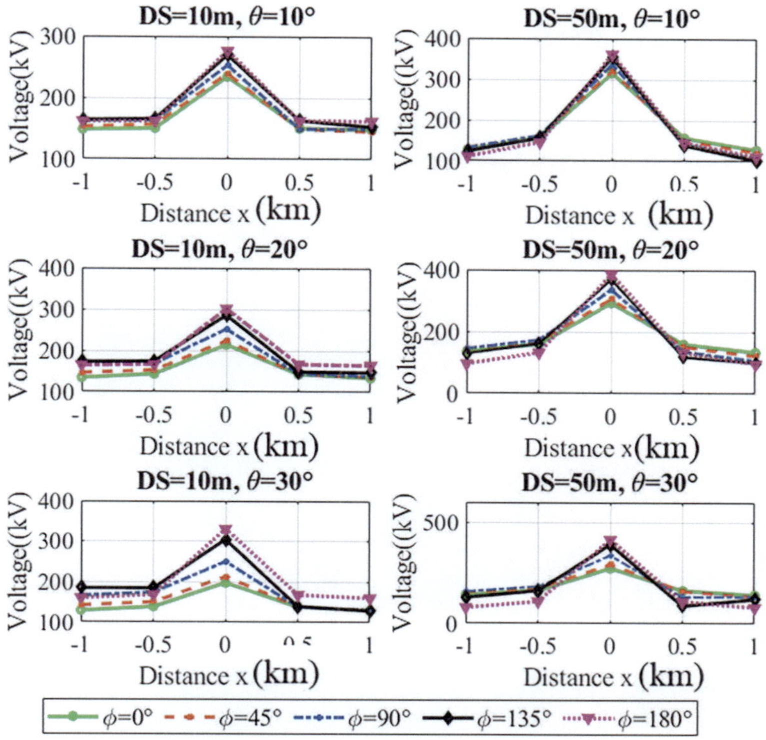

Fig. 5.8 The effect of lightning channel inclination on the peak of the induced voltage on the line due to a ground stroke

5.5.2 *The Effect of the Strike of the Inclined Stroke Channel to the Ocean Surface on the Induced Voltage on the Overhead Line*

Figure 5.11 shows the induced voltage waveshape for the inclined stroke channel at $r = 0$ m. Figure 5.12 shows the peak of the induced voltage at $r = 0$ m. For equal inclination angle (equal θ and φ), by increasing the distance between the overhead line and the coastline, the peak of the induced voltage increases.

For equal θ if $0° < \varphi < 90°$, by increasing φ, the peak of the induced voltage decreases. At $\varphi = 90$, increasing θ does not have such an influence on the peak of the induced voltage. For $90° < \varphi < 180°$ by increasing θ, the peak of the induced voltage increases.

The same as striking to ground the maximum of the peak at the middle of the line occurs when the channel angles are, $\varphi = 180°$ and $\theta = 30°$.

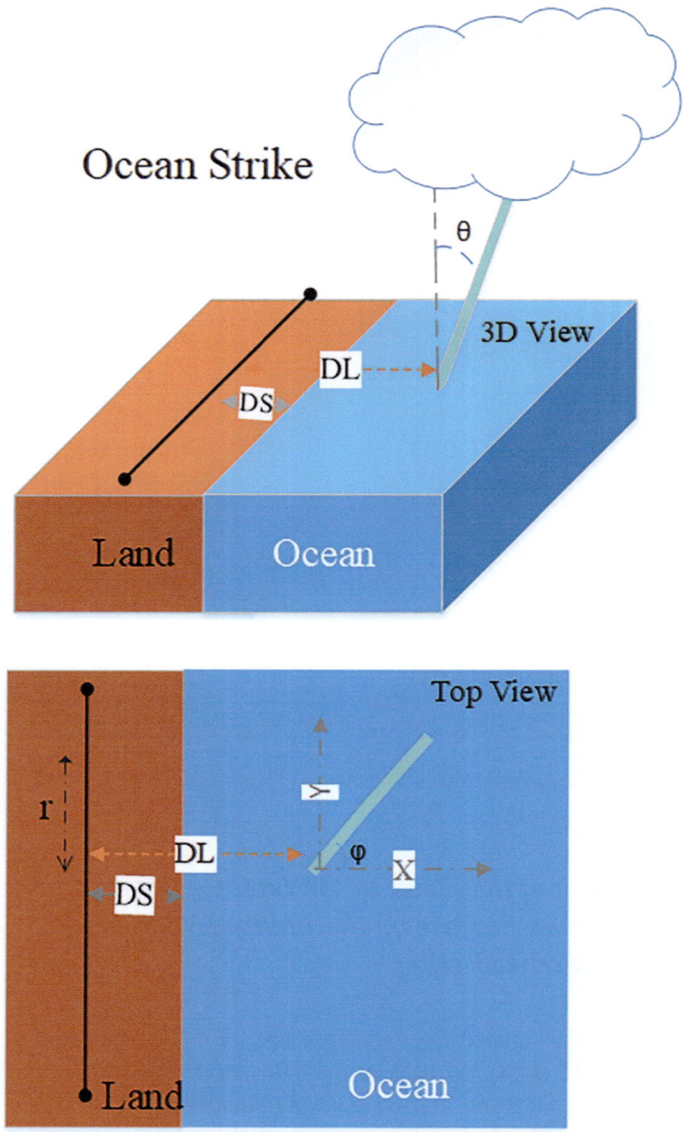

Fig. 5.9 The geometry of configuration

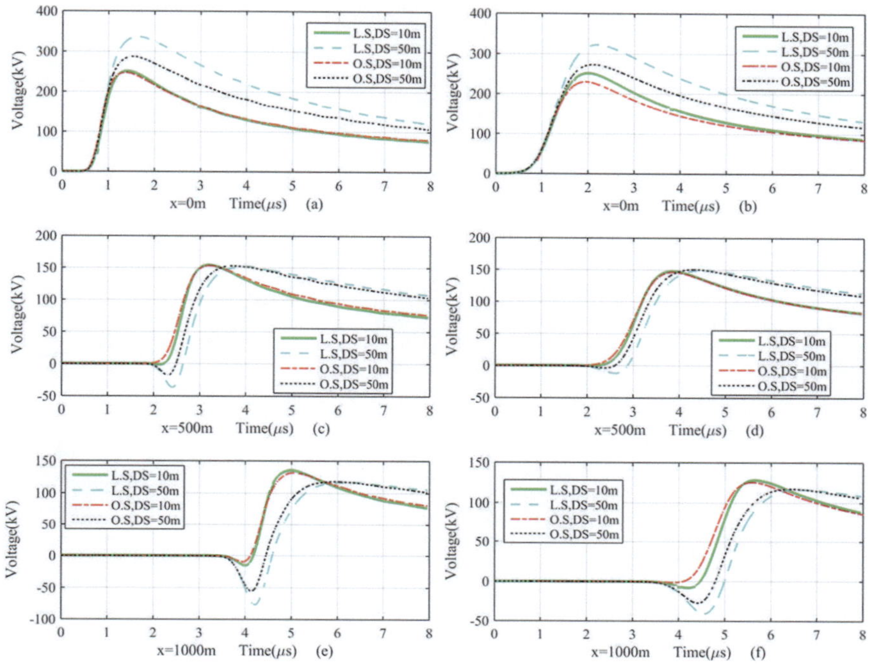

Fig. 5.10 The effect of striking point (ground or ocean) on the induced voltage of vertical lightning channel on overhead line ($\theta = 0°$, $\varphi = 0$). **(a)** $r = 0$ m, 0.25/100 μs, **(b)** $r = 0$ m, 1/200 μs, **(c)** $r =$ 500 m, 0.25/100 μs, **(d)** $r = 500$ m, 1/200 μs, **(e)** $r = 1000$ m, 0.25/100 μs, **(f)** $r = 1000$ m, 1/200 μs (L.S = land side, O.S = ocean side)

5.5.3 The Amplitude of Induced Voltage Along the Overhead Line Due to Strike of Inclined Stroke Channel to the Ocean Surface

Figure 5.13 shows the amplitude of induced voltage along the overhead line due to strike of inclined stroke channel to the ocean surface.

5.5.4 Comparison Between Results of FEM and FDTD

In order to evaluate the results of the FEM method, they are compared with the results of the FDTD method in Fig. 5.14 for three horizontal angles ($\varphi = 0°$, 90°, 180°) and a vertical angle $\theta = 30°$. In this simulation DL = 10 m and DS = 50 m. As can be seen, the results of the FEM and FDTD methods are close to each other.

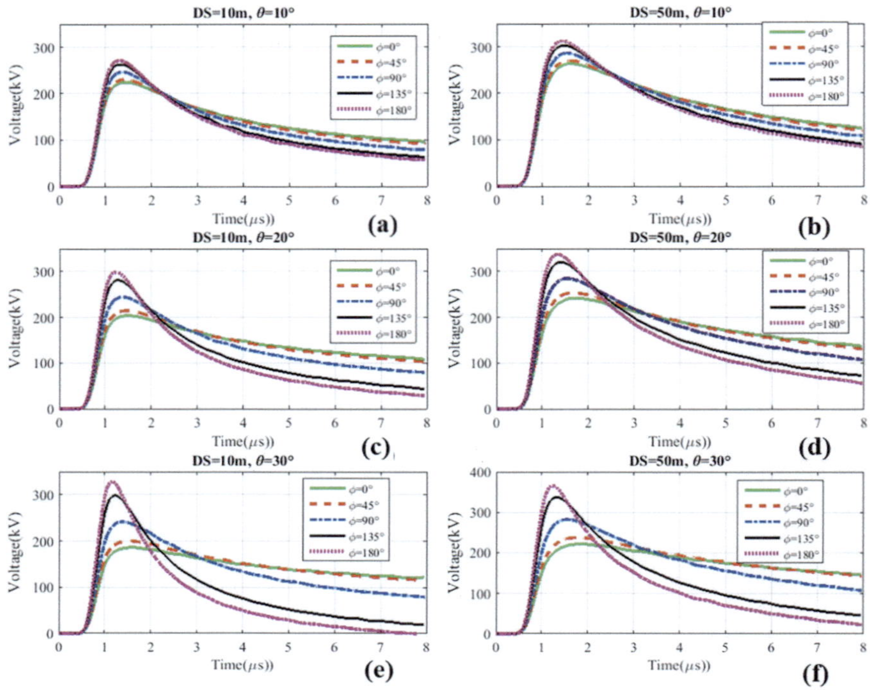

Fig. 5.11 The effect of slope angle of return stroke channel on the induced voltage at the middle of the line due to the stroke to the ocean surface for 0.25/100 μs. **(a)** $\theta = 10°$ and DS = 10 m, **(b)** DS = 50 m and $\theta = 10°$, **(c)** DS = 10 m and $\theta = 20°$, **(d)** DS = 50 m and $\theta = 20°$, **(e)** DS = 10 m and $\theta = 30°$, **(f)** DS = 50 m and $\theta = 30°$

5.5.5 The Effect of Lightning Current Waveshape on Induced Voltage

The effect of lightning current waveshape on the amplitude of induced voltage at the middle of the line for inclined stroke channel with current waveshapes of 0.25/100 μs and 1/200 μs is shown in Fig. 5.15. In this figure, the maximum and minimum amplitudes of the induced voltage are shown for each current waveshape, and the red line shows the median of the induced voltage. The upper and lower sides of the rectangular bar chart show the 25% and 75% range of induced voltage.

From Fig. 5.15, we can judge that the medians of the induced voltage for vertical and non-vertical stroke channels are close to each other.

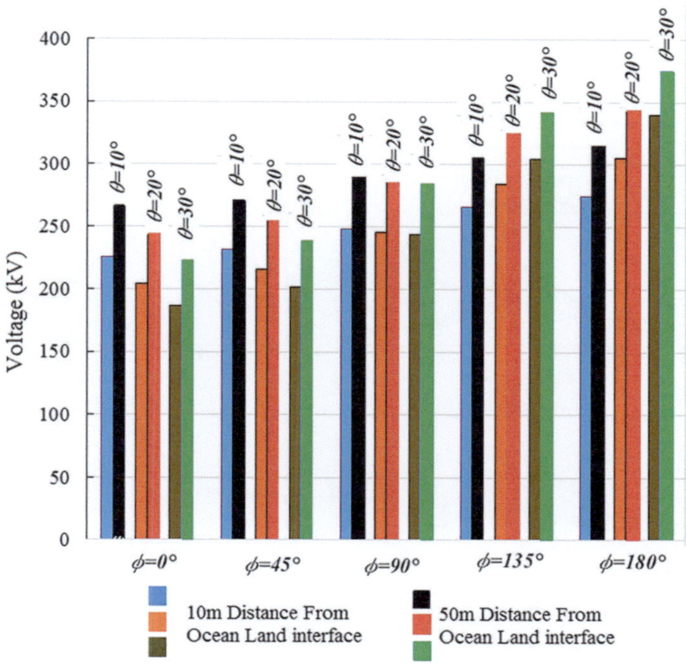

Fig. 5.12 The peak of induced voltage on middle of overhead line during stroke to ocean

5.6 The Effect of an Inclined Stroke Channel on the Induced Voltage on an Overhead Line in the Presence of the Ground with Two Horizontal Layers

In this section, the effect of an inclined stroke channel on the induced voltage on the overhead line is analyzed. The ground consists of two horizontal layers, and different parameters of the ground are considered.

5.6.1 The Modeling of Situation for the Ground with Two Horizontal Layers

COMSOL software is used to model the RF module [28]. This module, using 3D FEM, can solve the electromagnetic equations of lightning in time and frequency domains, and the time domain of electromagnetic wave (TEMW) of the RF module is used. The schematic configuration of the simulation is shown in Fig. 5.16.

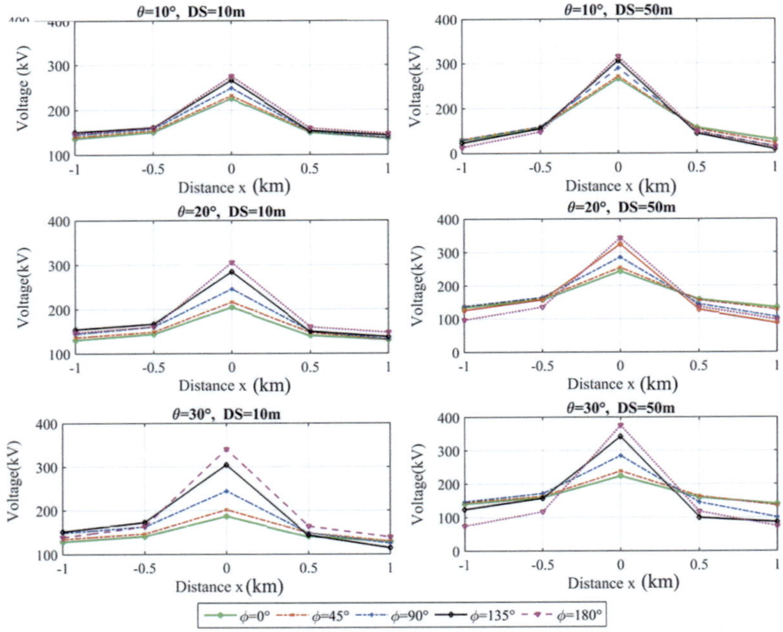

Fig. 5.13 The effect of stroke channel slope on the amplitude of induced voltage along the line due to the strike of lightning to the ocean surface

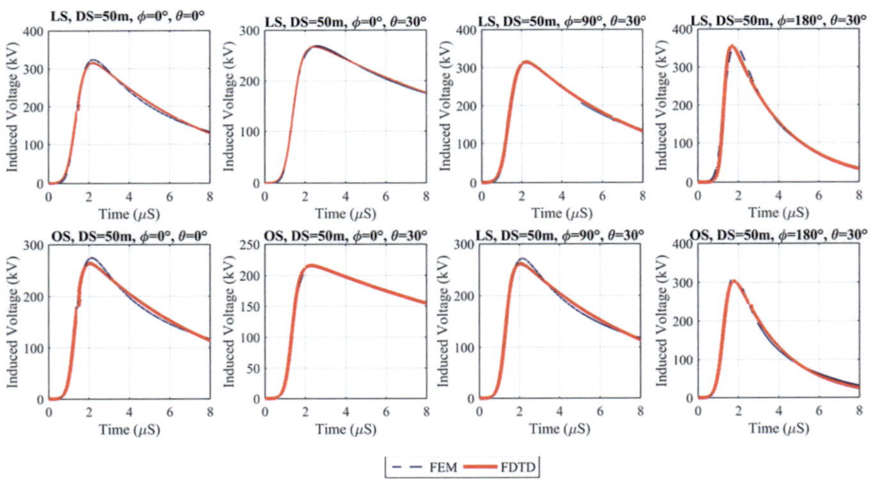

Fig. 5.14 The comparison between FEM and FDTD results for: $r = 0$ m, DL = 10 m, lightning waveshape=1/100 μs (L.S = land stroke, O.S = ocean stroke)

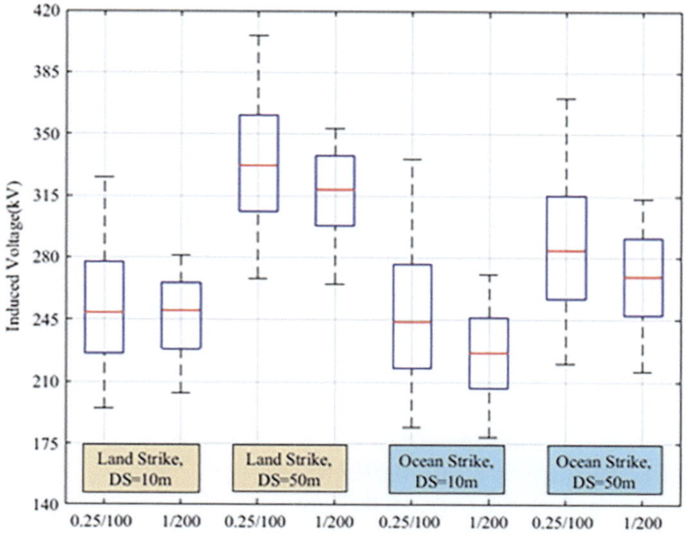

Fig. 5.15 The effect of lightning current waveshape on induced voltage in the presence of mixed ocean-land trace

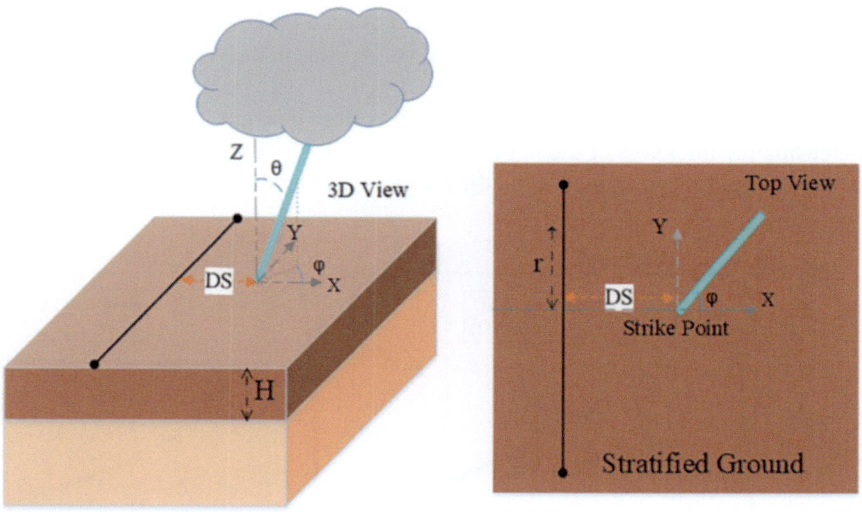

Fig. 5.16 The schematic configuration of the simulation (σ_u = conductivity of upper layer, σ_d = conductivity of deeper layer)

5.6.2 The Results of Simulation for Ground with Two Horizontal Layers for Vertical Stroke Channel

The geometric configuration of the simulation is shown in Fig. 5.16. Figure 5.17 shows the comparison of results for 3D FEM and results of FDTD (which are from Ref. [25]).

As can be seen in Fig. 5.17, the results of these two methods are close to each other.

In order to analyze the effect of two layers with different conductivities and current waveshapes, simulation results are shown in Fig. 5.18. In the following simulation results:

$r = 0, 500, 1000$ m
Stroke channel is vertical ($\theta = 0°$, $\varphi = 0°$)
Upper layer conductivity (σ_u) $= 0.01$S/m
Deeper layer conductivity (σ_d) $= 0.001$S/m
$H = 10$ m, 50 m, ∞
Waveshapes: 0.25/100 μs, 1/200 μs
Current amplitude = 50 kA.

The comparison of results shows the effect of each parameter.

The characteristic of the peak of induced voltage due to the vertical stroke channel and two-layer ground is shown in Fig. 5.19.

This figure shows how the peak of induced voltage changes along the line and by decreasing the thickness of the upper layer, the peak of induced voltage at far point decreases (because by reducing the thickness of layer with higher conductivity the

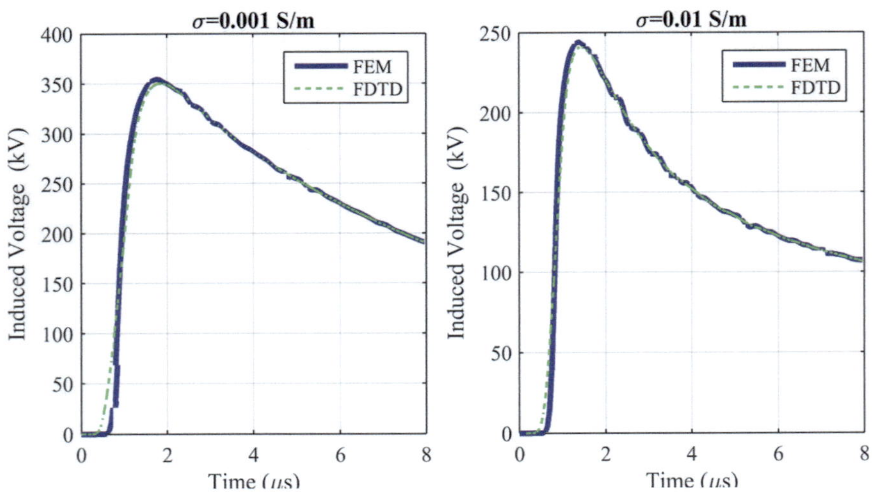

Fig. 5.17 The induced overvoltage on a single-conductor overhead line due to a vertical stroke channel is evaluated by FEM and FDTD methods

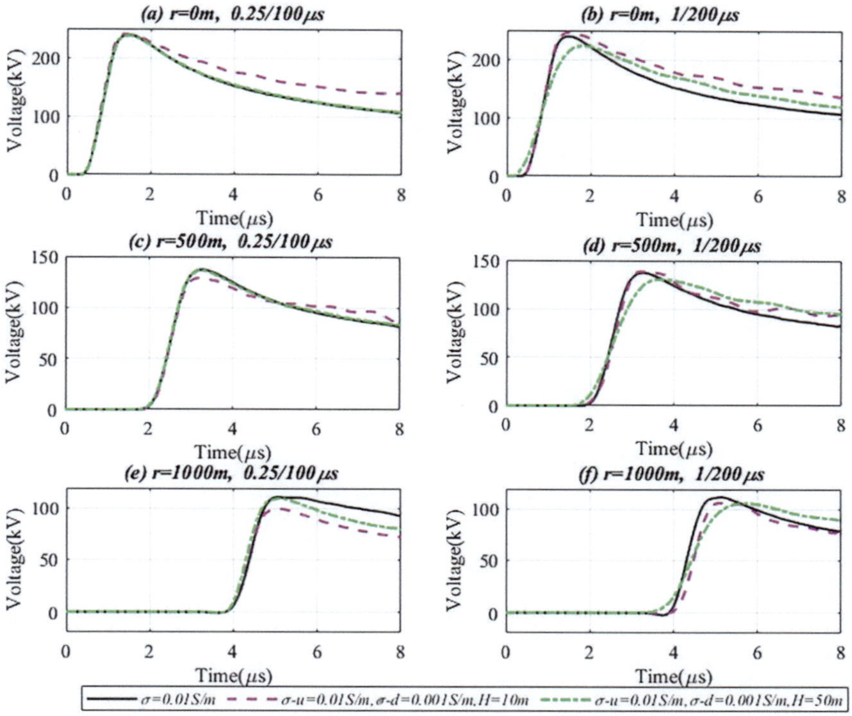

Fig. 5.18 The effect of a two-layer ground with different conductivities and current waveshapes on the induced voltage due to the vertical stroke channel compared with a single-layer ground (black line)

electromagnetic wave passes through low conductivity layer (higher losses) therefore peak of induced voltage at far point reduces).

5.6.3 The Results of Simulation for Ground with Two Horizontal Layers for Inclined Stroke Channel

The waveshapes of induced overvoltage due to a 1/200 μs lightning current which strikes the ground in the middle of the line (DS = 100 m, Fig. 5.16) can be seen in Fig. 5.20.

The peaks of induced voltages are shown in Fig. 5.21.

From these two figures, it can be deduced that, by increasing the horizontal angle (φ) when the vertical angle (θ) is constant, the peak of induced voltage increases, and the waveshape rise and fall times are faster. For channels that $0° < \varphi < 90°$, by increasing θ, the induced voltage peak decreases.

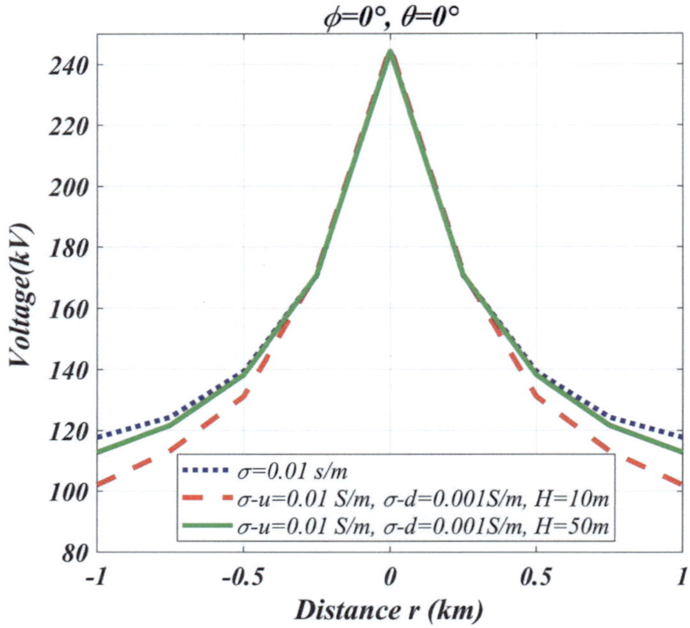

Fig. 5.19 The characteristic of the peak of the induced voltage due to the vertical stroke channel and the two-layer ground

Figure 5.22 shows the induced lightning overvoltage's profile along the overhead line. By comparing the profiles of induced overvoltages we can judge that when the distance between observation point and middle of the line increases the peak of overvoltage in the presence of a single layer ground is more than in the case of a double layer ground.

Comparison of results for $H = 10$ m with $H = 50$ m an $H = \infty$ shows that for an upper layer with higher thickness the peak on induced voltage is higher because the conductivity of the upper layer is higher than the deeper layer.

5.6.4 The Results of Simulation for Ground with Two Horizontal Layers When the Deeper Layer Conductivity is Higher Than the Upper Layer Conductivity

For this case, the conductivity of the upper layer is 0.001 S/m and the conductivity of the deeper layer is 0.01 S/m.

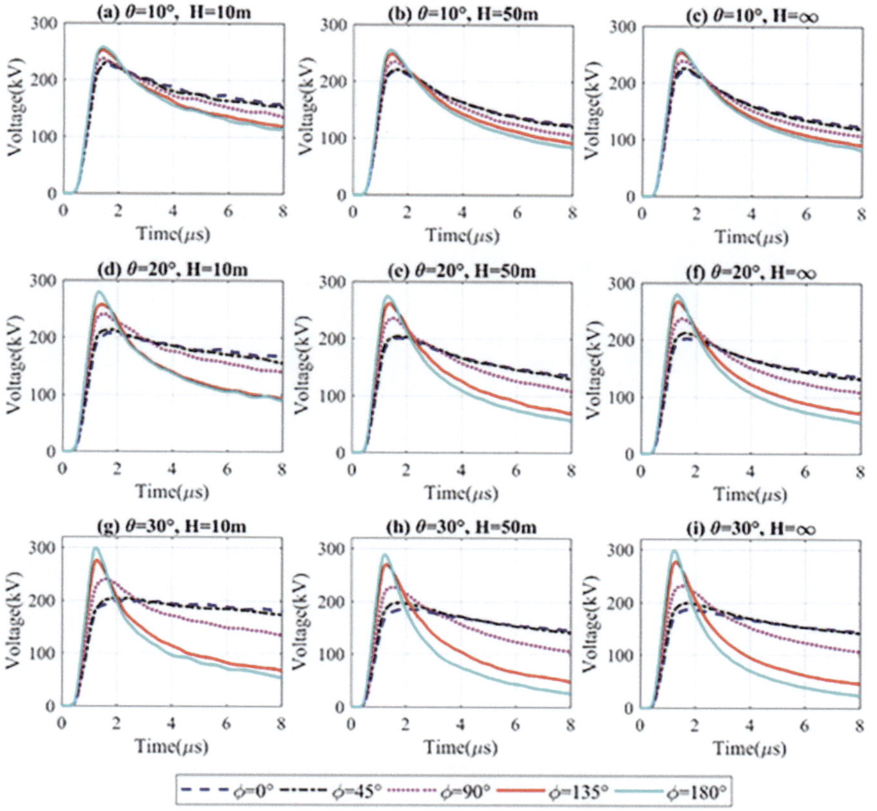

Fig. 5.20 The waveshapes of induced voltages are due to the inclined stroke channel for $r = 0$ m (conductivity of upper layers = 0.01 S/m, conductivity of deeper layer = 0.001 S/m)

5.6.4.1 The Induced Voltage Due to Vertical Stroke Channel

The waveshapes of induced overvoltages are shown in Fig. 5.23. The observation points are at $r = 0$, 500, 1000 m, and the deeper layer conductivity is higher than the upper layer conductivity. The current waveshapes are 0.25/100 μs and 1/200 μs. The geometrical configuration of the simulation is as shown in Fig. 5.16 (for this case, $\varphi = 0°$, $\theta = 0°$, and DS = 100 m). As can be seen in Fig. 5.23, when the thickness of the upper layer reduces, the peak of the induced voltage at the middle of the line decreases.

The profile of induced overvoltage along the line is shown in Fig. 5.24. According to this figure, by increasing the distance between the observation point and the middle of the line, the peaks of overvoltages for $H = 50$ m and ∞ is reduced more than the peak of overvoltage for $H = 10$ m.

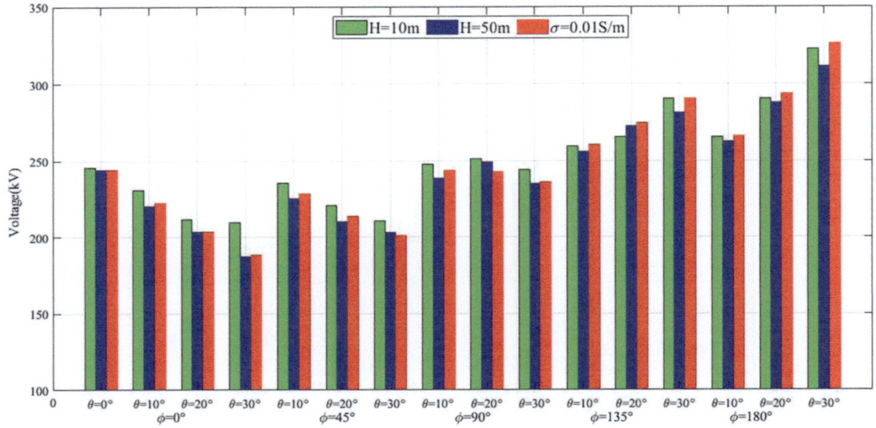

Fig. 5.21 The effect of return stroke channel angle and height of upper layer of ground on the peak of induced voltage (conductivity of upper layers = 0.01 S/m, conductivity of deeper layer = 0.001 S/m)

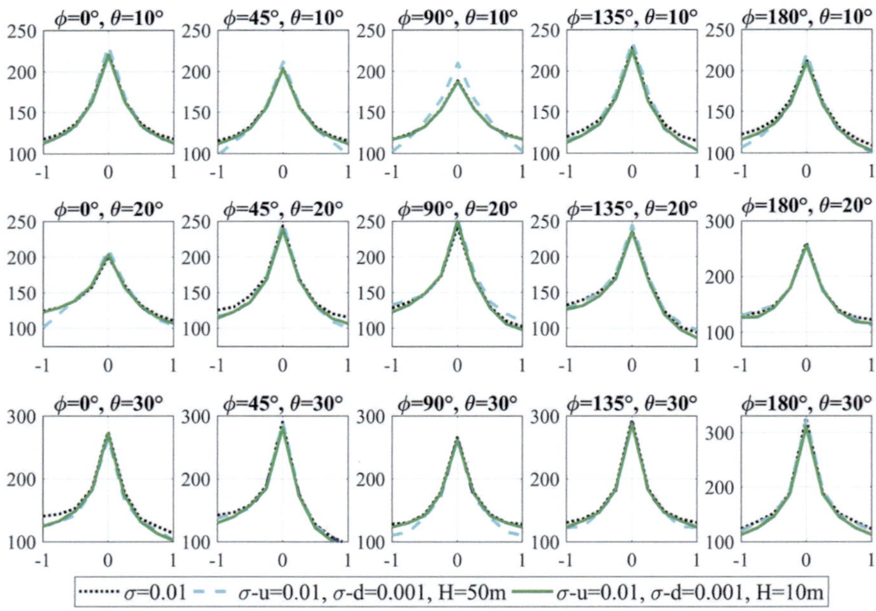

Fig. 5.22 The peak of induced lighting overvoltage (kV) along a single overhead line due to an inclined return stroke channel on the ground with different layer thicknesses

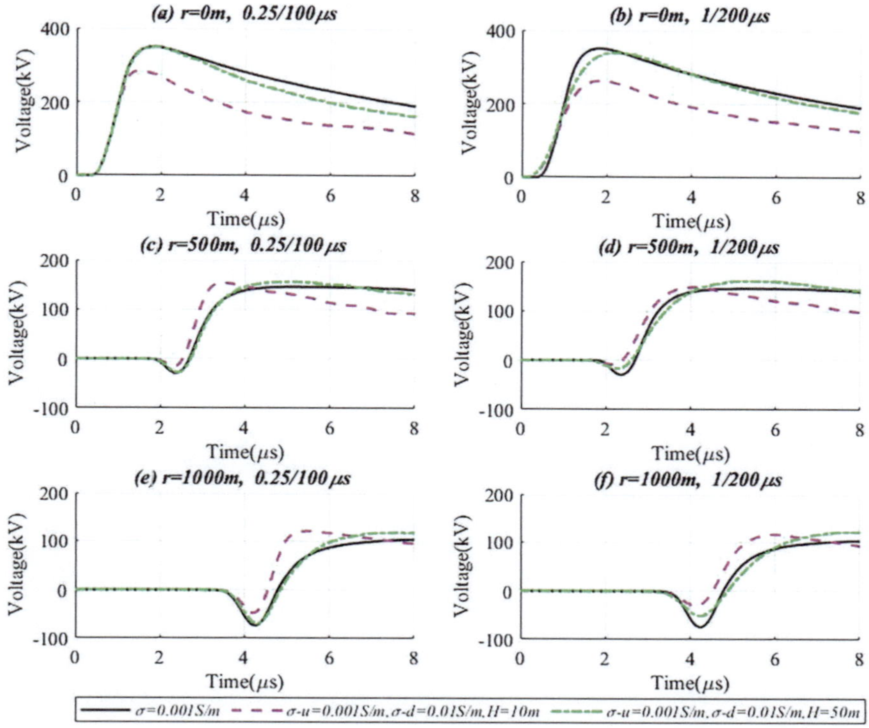

Fig. 5.23 The effect of the thickness of the upper layer and the current waveshape on the induced voltage due to the vertical stroke channel compared with the single-layer ground (black color line)

5.6.4.2 The Induced Voltage Due to Inclined Stroke Channel

Results of the simulation for a two-layer ground when the conductivity of the upper layer is less than the conductivity of the deeper layer are shown in Fig. 5.25.

In this figure, the waveshapes of induced overvoltages at the middle of the line are shown for different inclination angles and thicknesses of the upper layer for the current waveshape of 0.25/100 μs. By comparing the waveshapes, we can judge that the horizontal angle of the channel (φ) has a significant effect on the rise and fall times of the induced overvoltage waveshape. For the same vertical angle (θ), by increasing the horizontal angle from 0° to 180°, the rise and fall times of induced overvoltage decrease significantly.

Figure 5.26 shows the peak of induced overvoltage at the middle of the line. From Fig. 5.26, it can be deduced that the peak of induced overvoltage decreases when $0° < \varphi < 90°$ and θ increases. At $\varphi = 90°$, variation of θ does not have an important effect on the peaks of induced overvoltages, for $90° < \varphi < 180°$, by increasing θ, the peak of the induced overvoltage increases.

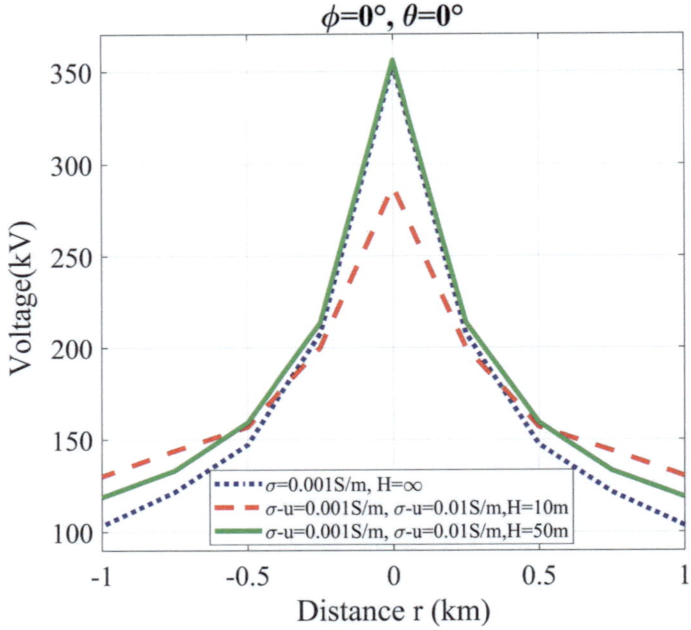

Fig. 5.24 The profile of induced overvoltage due to vertical return stroke

Figure 5.27 shows the profile of the peak of induced overvoltage along the whole line when the upper layer conductivity is less than the conductivity of the deeper layer.

5.6.4.3 The Effect of Current Waveshape at Base of Channel on Induced Overvoltage on Overhead Line

The effects of the current waveshape and the inclined angle of the return stroke channel on the induced voltage at the middle of the line in the presence of multiple horizontal ground layers are investigated in this section.

For a ground with two horizontal layers, when the conductivities of the two layers are different, electrical waves are reflected from the boundary between the two layers. The reflected waves disappear after several reflections, but they affect the induced voltage. For a single-layer ground, we do not have these reflections, and the induced voltages for two different current waveshapes (0.25/100 μs, 1/200 μs) are close to each other.

Figure 5.28 shows the statistical results for the peak of induced voltage for two different current waveshapes.

In Fig. 5.28, the minimum and maximum of induced voltages are shown for each case, and red lines show the median values for each case. The break lines show the range of variation for the peak of the induced voltage of each case.

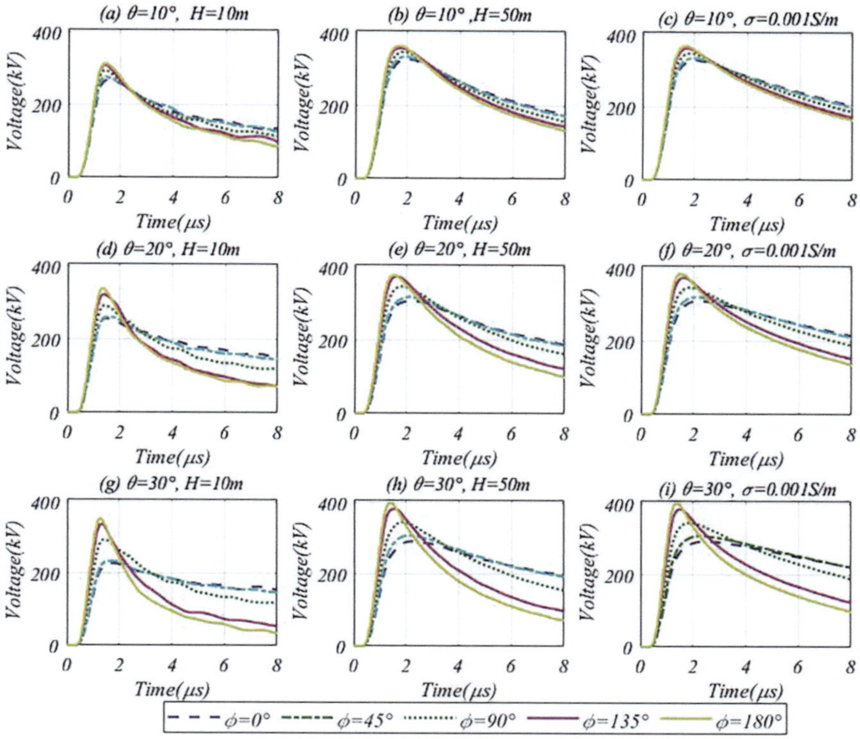

Fig. 5.25 The waveshapes of induced overvoltage at the middle of the line due to inclined stroke channel when the conductivity of the upper layer is 0.001 S/m and the conductivity of the deeper layer is 0.01 S/m, **(c)**, **(f)**, and **(i)** graphs are for a single-layer ground

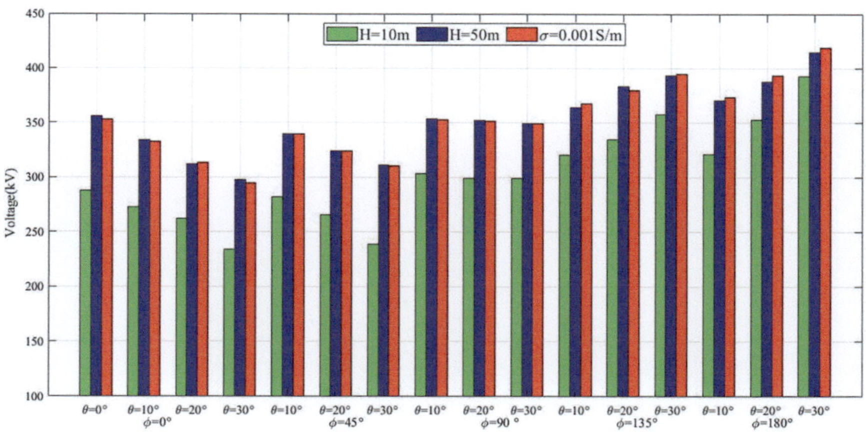

Fig. 5.26 The effect of return stroke channel angle and thickness of the upper layer of ground on the peak of induced overvoltage when the conductivity of the upper layer is 0.001 S/m and the conductivity of the deeper layer is 0.01 S/m, compared with single-layer ground (red color bar)

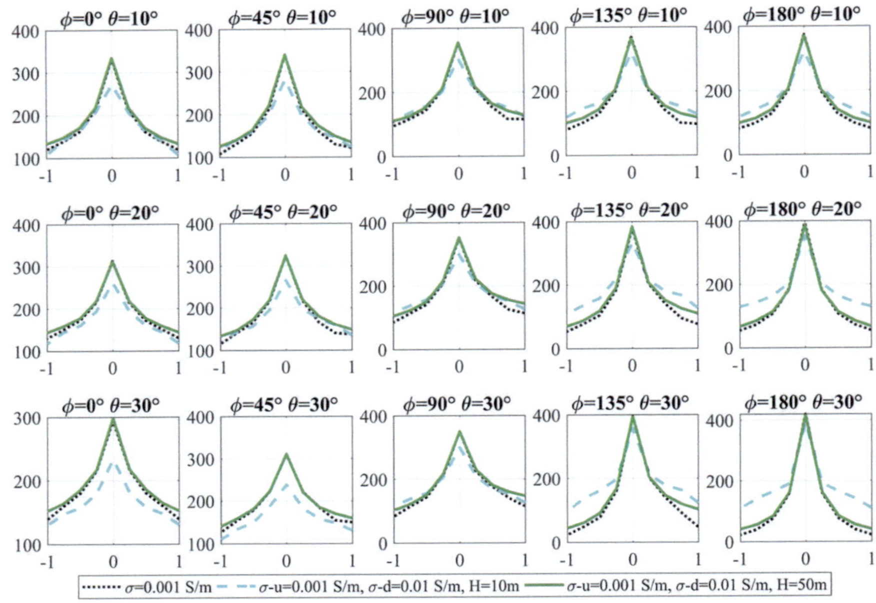

Fig. 5.27 The profile of the peak of induced overvoltage (kV) along the single conductor line due to inclined return stroke channel when the conductivity of the upper layer is 0.001 S/m and the conductivity of the deeper layer is 0.01 S/m, compared with the single-layer ground (black dot line)

By comparing the results, we can judge that the peak and median value of induced voltage for two different current waveshapes are similar, but for the ground with two layers, the induced voltage is affected by the current waveshape.

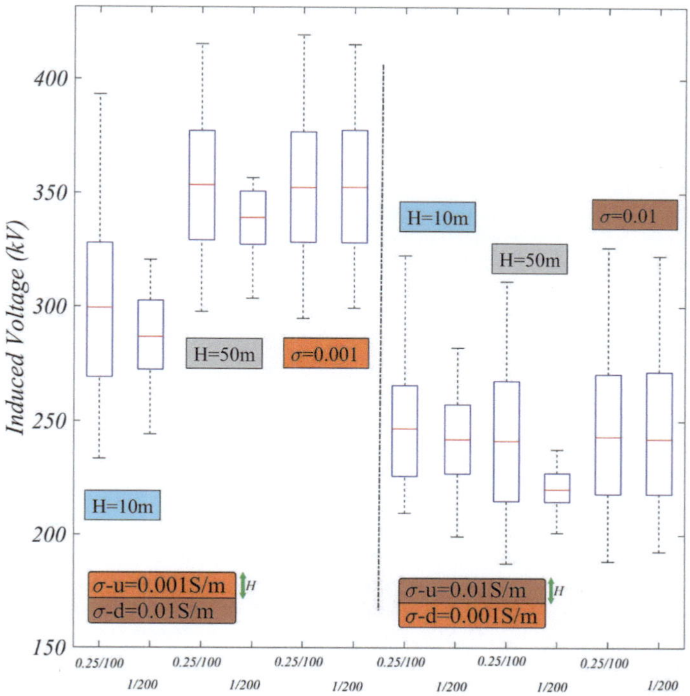

Fig. 5.28 The statistical results for the peak of induced voltage for two different current waveshapes (0.25/100 μs, 1/200 μs), compared with single-layer ground ($\sigma = 0.001$ S/m, and $\sigma = 0.01$ S/m)

References

1. A. Andreotti, C. Petrarca, A. Pierno, On the effects of channel tortuosity in lightning-induced voltages assessment. IEEE Trans. Electromagn. Compat. **57**, 1096–1102 (2015)
2. A. Borghetti, A. Morched, F. Napolitano, C.A. Nucci, M. Paolone, Lightning-induced overvoltages transferred through distribution power transformers. IEEE Trans. Power Delivery **24**, 360–372 (2008)
3. A. Piantini, Analysis of the effectiveness of shield wires in mitigating lightning-induced voltages on power distribution lines. Electr. Power Syst. Res. **159**, 9–16 (2018)
4. A. Andreotti, F. Mottola, A. Pierno, D. Proto, On the statistical characterization of lightning-induced voltages. Appl. Sci. **8**(651), 1–15 (2018)
5. A. Andreotti, A. Piantini, A. Pierno, R. Rizzo, Lightning-induced voltages on complex power systems by using CiLIV: the effects of channel tortuosity. IEEE Trans. Power Delivery **33**, 680–688 (2017)
6. A. Borghetti, J.A. Gutierrez, C.A. Nucci, M. Paolone, E. Petrache, F. Rachidi, Lightning-induced voltages on complex distribution systems: models, advanced software tools and experimental validation. J. Electrostat. **60**, 163–174 (2004)
7. E. Perez, A. Delgadillo, D. Urrutia, H. Torres, Optimizing the surge arresters' location for improving lightning induced voltage performance of distribution network, in *2007 IEEE Power Engineering Society General Meeting*, pp. 1–6 (2007)
8. I. Matsubara, S. Sekioka, Analytical formulas for induced surges on a long overhead line caused by lightning with an arbitrary channel inclination. IEEE Trans. Electromagn. Compat. **51**, 733–740 (2009)

9. F. Napolitano, A. Borghetti, C.A. Nucci, F. Rachidi, M. Paolone, Use of the full-wave finite element method for the numerical electromagnetic analysis of LEMP and its coupling to overhead lines. Electr. Power Syst. Res. **94**, 24–29 (2013)
10. M. Akbari, K. Sheshyekani, A. Pirayesh, F. Rachidi, M. Paolone, A. Borghetti, C.A. Nucci, Evaluation of lightning electromagnetic fields and their induced voltages on overhead lines considering the frequency dependence of soil electrical parameters. IEEE Trans. Electromagn. Compat. **55**, 1210–1219 (2013)
11. J. Paknahad, K. Sheshyekani, M. Hamzeh, D. Li, F. Rachidi, The influence of the slope angle of the ocean–land mixed propagation path on the lightning electromagnetic fields. IEEE Trans. Electromagn. Compat. **57**, 1086–1095 (2015)
12. J. Paknahad, K. Sheshyekani, F. Rachidi, M. Paolone, Lightning electromagnetic fields and their induced currents on buried cables. Part II: The effect of a horizontally stratified ground. IEEE Trans. Electromagn. Compat. **56**, 1146–1154 (2014)
13. K. Sheshyekani, J. Paknahad, Lightning electromagnetic fields and their induced voltages on overhead lines: the effect of a horizontally stratified ground. IEEE Trans. Power Delivery **30**, 290–298 (2015)
14. M.E.M. Rizk, F. Mahmood, M. Lehtonen, E.A. Badran, M.H. Abdel-Rahman, Investigation of lightning electromagnetic fields on underground cables in wind farms. IEEE Trans. Electromagn. Compat. **58**, 143–152 (2016)
15. W. Li, T. Xie, W. Zhang, Z. Zhang, D. Mei, Calculation method and influencing factors of lightning-induced voltage on overhead distribution line, in *2017 IEEE 3rd Information Technology and Mechatronics Engineering Conference* (ITOEC), pp. 420–425 (2017)
16. H. Chen, Y. Du, M. Yuan, Q.H. Liu, Lightning-induced voltages on a distribution line with surge arresters using a hybrid FDTD–SPICE method. IEEE Trans. Power Delivery **33**, 2354–2363 (2018)
17. T.H. Thang, Y. Baba, V.A. Rakov, A. Piantini, FDTD computation of lightning-induced voltages on multiconductor lines with surge arresters and pole transformers. IEEE Trans. Electromagn. Compat. **57**, 442–447 (2015)
18. M. Azadifar, P. Dehkhoda, S.H.H. Sadeghi, R. Moini, A hybrid FDFD–MoM technique for susceptibility evaluation of a transmission line inside a perforated enclosure. IEEE Trans. Electromagn. Compat. **56**, 1474–1479 (2014)
19. M.E.M. Rizk, F. Mahmood, M. Lehtonen, E.A. Badran, M.H. Abdel-Rahman, Induced voltages on overhead line by return strokes to grounded wind tower considering horizontally stratified ground. IEEE Trans. Electromagn. Compatibility **58**, 1728–1738 (2016)
20. Q. Li, M. Rubinstein, J. Wang, L. Cai, M. Zhou, Y. Fan, F. Rachidi, On the influence of the soil stratification and frequency-dependent parameters on lightning electromagnetic fields. Electr. Power Syst. Res. **178**, 1–10 (2020)
21. R. Moini, S.H.H. Sadeghi, B. Kordi, F. Rachidi, An antenna-theory approach for modeling inclined lightning return stroke channels. Electr. Power Syst. Res. **76**, 945–952 (2006)
22. A.F. Nematollahi, B. Vahidi, The effect of the inclined lightning channel on electromagnetic fields and the induced voltages on overhead lines. Electr. Eng. **103**, 3163–3176 (2021)
23. M. Natsui, A. Ametani, J. Mahseredjian, S. Sekioka, K. Yamamoto, FDTD analysis of distribution line voltages induced by inclined lightning channel. Electr. Power Syst. Res. **160**, 450–456 (2018)
24. R. Khosravi, S.H.H. Sadeghi, R. Moini, Electromagnetic field due to lightning strike to a tall tower sitting on a mountainous terrain. IEEE Trans. Electromagn. Compat. **58**, 1090–1099 (2016)
25. M. Natsui, A. Ametani, J. Mahseredjian, S. Sekioka, K. Yamamoto, 3-D FDTD analysis of lightning-induced voltages in distribution lines due to inclined lightning. IEEE Trans. Electromagn. Compat. **63**, 189–197 (2020)
26. A.F. Nematollahi, Invesigation of inducved overvoltage caused by lightning in distribution networks considering environmental and soil conditions. Ph.D. thesis, Amirkabir University of Technology, Tehran, Iran (2023)

27. A.F. Nematollahi, Analysis of performance of overhead lines against lightning induced voltage in presence of different ground structres. Post Doctoral Report, Amirkabir University of Technology, Tehran, Iran (2024)
28. COMSOL, *COMSOL RF Module User's Guide* (COMSOL, Inc., Stockholm, Sweden, 2016)
29. A.F. Nematolahi, B. Vahidi, The efect of the inclined lightning channel on electromagnetic fields and the induced voltages on overhead lines. Electr. Eng. **103**, 3163–3176 (2021)
30. IEC, *Protection Against Lightning Part 1: General Principles* (IEC, 2011)
31. M.R.B. Tavakoli, B. Vahidi, Transmission-lines shielding failure-rate calculation by means of 3-D leader progression models. IEEE Trans. Power Delivery **26**, 507–516 (2011)
32. M. Yahyaabadi, B. Vahidi, M.R.B. Tavakoli, Estimation of shielding failure number of different configurations of double-circuit transmission lines using leader progression analysis model. Electr. Eng. **92**, 79–85 (2010)
33. Sheshyekani, J. Paknahad, The effect of an ocean-land mixed propagation path on the lightning electromagnetic fields and their induced voltages on overhead lines. IEEE Trans. Power Deliv. **30**, 229–236 (2015)

Chapter 6
Flashover Prediction in Distribution Lines

Abstract This chapter presents a comprehensive methodology for predicting indirect lightning-induced flashover that incorporates two crucial elements often overlooked in simplified approaches: the dynamic behavior of insulators under voltage stress (using the volt-time disruptive model) and the complex influence of non-uniform ground conditions. Specifically, this chapter will:

1. Review and critique existing flashover prediction methods
2. Introduce the volt-time disruptive model
3. Detail the key factors influencing induced voltages
4. Describe advanced modeling techniques
5. Present and validate new analytical formulas
6. Overview mitigation strategies.

6.1 Introduction

Lightning strikes, as comprehensively discussed in the preceding chapters, represent a major threat to the reliability and safety of power distribution systems. Chapter 1 elucidated the fundamental physics of lightning, covering its formation, types, and key characteristics, such as the stepped leader, return stroke, and dart leader. Chapter 2 detailed the various ways lightning interacts with distribution networks, highlighting the damaging effects of both direct and indirect strikes, and introduced basic protection concepts like surge arresters. Chapter 3 provided a thorough review of existing methods for calculating lightning-induced voltages, encompassing both analytical and numerical approaches, and discussing various return stroke channel models and soil models. It introduced the complexities involved in accurate voltage calculation. Chapter 4 focused on methods for identifying lightning channel-base current characteristics and determining appropriate parameters for the Heidler function, a crucial component in modeling the lightning current waveform. Finally, Chap. 5 investigated the specific influence of inclined return-stroke channels on induced voltages. Building upon this foundation, this chapter focuses on indirect lightning-induced

B. Vahidi and A. F. Nematolahi, *Lightning Induced Overvoltage and Current on Distribution Power Systems*, https://doi.org/10.1007/978-981-95-3712-9_6

flashover in overhead distribution lines, emphasizing advanced modeling techniques for accurate prediction and informed mitigation strategies.

The increasing reliance on sensitive electronic devices in modern society, coupled with growing concerns about climate change and its potential impact on lightning frequency [1], makes accurate flashover prediction more critical than ever. Furthermore, aging infrastructure in many regions necessitates a reevaluation of existing lightning protection paradigms. A 12% global increase in lightning strikes since 2000 has been attributed to rising atmospheric temperatures and increased convective activity [1]. This trend, combined with the widespread use of sensitive electronics and the aging of existing power infrastructure, highlights the urgency of improving lightning protection strategies.

This chapter presents a comprehensive methodology for predicting indirect lightning-induced flashover that incorporates two crucial elements often overlooked in simplified approaches: the dynamic behavior of insulators under voltage stress (using the volt-time disruptive model) and the complex influence of non-uniform ground conditions. Specifically, this chapter will:

1. Review and critique existing flashover prediction methods: Highlighting the limitations of the commonly used 1.5CFO criterion (Sect. 6.2).
2. Introduce the volt-time disruptive model: Explaining its principles and advantages as a more accurate and physically realistic approach to flashover prediction (Sect. 6.2).
3. Detail the key factors influencing induced voltages: With a particular emphasis on the complex effects of soil resistivity, ground stratification, and the presence of ocean-land interfaces (Sect. 6.3).
4. Describe advanced modeling techniques: Explaining the 3D Finite Element Method (FEM) simulations and Monte Carlo analysis used for accurate voltage calculation and statistical analysis (Sect. 6.4).
5. Present and validate new analytical formulas: Providing simplified formulas for estimating annual flashover rates in scenarios with horizontally stratified soil and mixed ocean-land propagation paths (Sect. 6.5).
6. Overview mitigation strategies: Discussing the protection methods (Sect. 6.6).
7. Summarize some case studies: Summarizing previous sections by case studies (Sect. 6.7).

By integrating these elements, this chapter aims to provide researchers and practicing power engineers with the knowledge and tools necessary to design more resilient and reliable distribution systems, minimizing the risk of lightning-induced flashovers and ensuring the continuity of power supply.

6.2 Review of Flashover Prediction Methods

Accurately predicting the occurrence of flashover on overhead distribution lines due to indirect lightning strikes is a complex task, involving the interplay of numerous factors related to the lightning discharge itself, the characteristics of the overhead line, and the surrounding environment. As discussed in Chap. 3, various methods have been developed to calculate lightning-induced voltages, ranging from simplified analytical approaches to sophisticated numerical techniques. However, regardless of the method used to calculate the induced voltage, a fundamental component of any flashover prediction methodology is the criterion used to determine whether a given voltage waveform will, in fact, cause an insulator breakdown. This section reviews the two primary approaches used in the literature: the widely adopted, but simplified, 1.5CFO criterion, and the more physically realistic and accurate Volt-Time Disruptive Model. The historical development will be examined, including the underlying assumptions, advantages, and limitations of each method, ultimately demonstrating the need for a more advanced approach, particularly when dealing with complex scenarios.

6.2.1 The 1.5CFO Criterion: A Historical Perspective and Its Limitations

The 1.5CFO criterion has been a cornerstone of lightning protection design for many decades. Its simplicity and ease of application have made it a popular choice for engineers.

The criterion states that a flashover will occur if the *peak* value of the lightning-induced voltage on the line exceeds 1.5 times the insulator's Critical Flashover Voltage (CFO).

The CFO itself is typically determined through standardized laboratory tests, as described in IEC 60060-1 [2]. In these tests, insulators are subjected to impulse voltages with a defined waveshape, usually a 1.2/50 μs wave, intended to represent a standard lightning impulse. The CFO is defined as the voltage level at which the insulator has a 50% probability of flashover under these standardized conditions.

The 1.5 multiplication factor in the 1.5CFO criterion is an empirically derived value, intended as a safety margin to account for several factors that are *not* explicitly considered in the standard CFO test:

- **Non-standard Waveshapes**: Lightning-induced voltages in real-world scenarios rarely, if ever, perfectly match the standard 1.2/50 μs impulse. The waveshape is influenced by numerous factors, including the lightning current parameters (peak, rise time, tail time), the distance to the strike location, the ground conductivity, and the presence of nearby objects. The 1.5 factor was intended to provide a conservative estimate, encompassing a range of possible waveshapes. However, as research has shown, this factor is often insufficient for many realistic scenarios [3–10].

- **Insulator Aging and Degradation**: Over time, insulators can degrade due to environmental factors such as pollution, UV radiation, and mechanical stress. This degradation can reduce their withstand capability below the initially measured CFO. The 1.5 factor provides some margin for this degradation.
- **Statistical Variability**: Even under carefully controlled laboratory conditions, there is inherent statistical variability in the flashover voltage of insulators. The 1.5 factor attempts to account for this variability, ensuring a low probability of flashover.

Despite its widespread use, the 1.5CFO criterion suffers from several fundamental limitations that have been increasingly recognized in the literature:

- **Neglect of Waveshape Dynamics**: This is perhaps the most significant drawback. The criterion considers only the peak voltage, completely ignoring the shape and duration of the induced voltage waveform. This is a gross oversimplification of the physics of insulator breakdown. The time required for streamers and leaders to develop and bridge the gap is finite. A very fast-rising, short-duration pulse with a high peak might be less likely to cause a flashover than a slower-rising, longer-duration pulse with a lower peak, because the former might not provide sufficient time or energy for the breakdown process to complete.

  Numerous studies have shown that the front time (steepness) of the induced voltage has a significant impact on the flashover probability[11–15].
- **Oversimplified Insulator Model**: Insulators do not have a single, fixed breakdown voltage. Their response to transient overvoltage is a complex, dynamic process involving ionization, streamer formation, and leader propagation. The 1.5CFO criterion represents the insulator as a simple threshold, failing to capture this complex, time-dependent behavior.
- **Lack of Consideration for Ground Conditions**: The 1.5CFO criterion, in its basic form, does not explicitly account for the influence of ground conditions. Soil resistivity, ground stratification, and the presence of highly conductive bodies of water (such as the ocean) significantly affect the propagation of electromagnetic fields and, consequently, the magnitude and waveshape of induced voltages. The research presented in [5, 16–18] and further discussed in this chapter, demonstrates the crucial importance of considering these factors. Ignoring them can lead to substantial underestimation of flashover rates. Early work by Chowdhuri and Gross [19] recognized the influence of ground resistivity, but their model still relied on simplified representations.
- **Limited Applicability**: The studies [19–21], which introduce this criterion, assumed that the ground is a perfect conductor.

These limitations mean that relying solely on the 1.5CFO criterion can lead to under-designed lightning protection systems, resulting in higher-than-expected outage rates and increased risk of equipment damage. While it may be adequate for preliminary estimations or for systems with very high insulation levels, it is not suitable for situations where accurate flashover prediction is critical, particularly in areas with complex ground conditions or high lightning activity.

6.2.2 The Volt-Time Disruptive Model: A Physically Realistic Approach

The Volt-Time Disruptive Model offers a significantly more accurate and physically realistic approach to flashover prediction. This model, developed and refined over several decades [12, 21], recognizes that insulator breakdown is not solely determined by the peak voltage, but also by the time integral of the voltage waveform applied to the insulator. It accounts for the cumulative electrical stress experienced by the insulator and the dynamic nature of the breakdown process.

The core concept of the volt-time model is that the insulator's dielectric strength degrades over time when subjected to a voltage exceeding a certain threshold. This degradation is not instantaneous; it takes time for ionization, streamer formation, and leader propagation to occur. The longer the voltage exceeds the threshold, and the higher the voltage above that threshold, the greater the likelihood of flashover.

This concept is mathematically represented by the integration method, expressed by the following equation:

$$\mathrm{DE} = \int_{t_0}^{t} [U(t) - U_0]^K \mathrm{d}t, \quad t > t_0 \tag{6.1}$$

where

- DE: The Disruptive Effect. This is a dimensionless quantity that represents the accumulated stress on the insulator. It's a measure of how "close" the insulator is to flashing over. As DE increases, the probability of flashover increases.
- $U(t)$: The time-varying voltage waveform across the insulator (in kV). This is the induced voltage, calculated using the methods described in Sect. 6.4 (e.g., 3D FEM simulations).
- U_0: The onset voltage (in kV). This is the voltage level at which the insulator's dielectric strength begins to degrade, and the breakdown process initiates. Below U_0 The insulator is assumed to be able to withstand the voltage indefinitely.
- K: A dimensionless constant, typically close to 1, that accounts for the specific characteristics of the insulator material and geometry (e.g., its non-uniformity, the presence of defects, and the electrode configuration).
- t_0: The time (in μs) at which the voltage waveform $U(t)$ first exceeds the onset voltage U_0. The integration starts at this point.
- The integral is taken from t_0 to t.

The flashover criterion in the volt-time model is

Flashover occurs when DE reaches a critical value, DE*.

Crucially, U_0, K, and DE* are not universal constants. They are material and geometry-dependent parameters that must be determined experimentally through laboratory testing of the specific insulator type under consideration. Standard testing procedures, such as those described in IEC 60060-1 [2], are used to obtain these parameters. These tests typically involve applying various impulse voltage waveforms (not just the standard 1.2/50 μs wave) to the insulator and measuring the time to flashover. The resulting data is used to construct a volt-time curve, which plots the breakdown voltage as a function of the time to breakdown. The parameters U_0, K, and DE* are then derived from this curve, often using curve-fitting techniques.

The advantages of the Volt-Time Disruptive Model over the 1.5CFO criterion are substantial:

- **Increased Accuracy**: By considering the entire voltage waveform and the dynamic response of the insulator, the volt-time model provides a significantly more accurate prediction of flashover probability, especially for non-standard voltage waveforms.
- **Physical Basis**: The model is based on a more physically realistic understanding of the breakdown process, reflecting the cumulative stress on the insulator and the time-dependent nature of streamer and leader development.
- **Waveshape Dependence**: The model explicitly accounts for the shape and duration of the induced voltage, recognizing that these factors are crucial in determining flashover. This is a major improvement over the peak-voltage-only 1.5CFO criterion.
- **Flexibility**: The model can be applied to a wide range of insulator types and voltage waveforms, provided that the appropriate parameters (U_0, K, DE^*) are known.

Early work by Darveniza [12] laid the groundwork for the generalized integration method. Subsequent research by Hileman [21] and CIGRE Working Groups [22, 23] further developed and refined the model. More recently, researchers have combined the volt-time model with advanced numerical techniques, such as 3D FEM, to accurately calculate induced voltages and flashover probabilities under complex conditions [5, 16, 18, 24].

6.2.3 Comparative Studies: Demonstrating the Superiority of the Volt-Time Model

The improved accuracy of the Volt-Time Disruptive Model compared to the 1.5CFO criterion has been convincingly demonstrated through numerous studies, including the research presented in [5, 16, 18]. These studies employed detailed 3D FEM simulations, coupled with Monte Carlo analysis, to calculate lightning-induced voltages and flashover probabilities under various conditions, including:

- **Mixed Ocean-Land Propagation Paths**: This scenario is particularly relevant for coastal regions and offshore wind farms, where the high conductivity of seawater significantly alters the electromagnetic fields and induced voltages compared to a homogeneous land surface.
- **Horizontally Stratified Soil**: This scenario considers the common situation where the soil consists of two horizontal layers with different conductivities.

Figure 6.5 presents a schematic representation of a two-layer vertical ground model, highlighting the ocean-land interface.

The model emphasizes the stark contrast in electrical conductivity between seawater ($\sigma_{\text{ocean}} = 4$ S/m) and land-based soil ($\sigma_{\text{land}} = 0.001$ S/m). The parameter DS denotes the distance between the overhead transmission line and the ocean-land boundary. For comparative purposes, results for a single-layer ground model are also provided in Fig. 6.1. The results clearly demonstrate that the 1.5CFO criterion significantly underestimates the flashover rate, particularly as the distance from the coastline (DS) increases. For example, at a CFO of 200 kV, the underestimation ranges from 76 to 94% for different DS values. This substantial discrepancy is due to the simplified nature of the 1.5CFO criterion, which fails to capture the complex interactions between the lightning electromagnetic fields, the highly conductive seawater, and the overhead line, leading to altered voltage waveforms and increased flashover probability. The volt-time model, by contrast, accurately accounts for these effects, resulting in a more realistic prediction.

Similarly, Fig. 6.2 (adapted from [5]) shows a comparison for a scenario with horizontally stratified soil.

Figure 6.4 presents a schematic representation of a two-layer horizontal soil model. The upper layer has a finite depth (***H***) and conductivity ($\boldsymbol{\sigma_u}$), while the lower layer extends infinitely with conductivity ($\boldsymbol{\sigma_d}$). Two scenarios for soil conductivity are considered: one with $\sigma_u = 0.001$ S/m, $\sigma_d = 0.01$ S/m, and the other with $\sigma_u = 0.01$ S/m, $\sigma_d = 0.001$ S/m. For comparison, results from a single-layer ground model are also shown in Fig. 6.2.

Again, the 1.5CFO criterion consistently underestimates the flashover rate compared to the volt-time model. The degree of underestimation varies depending on the depth and conductivity of the soil layers, but it can be substantial, ranging from 60 to 85% in some of the cases presented. These results highlight the importance of considering both the soil stratification and the dynamic behavior of the insulator for accurate flashover prediction.

These examples, and numerous others presented in the literature [3, 5, 16, 18, 21, 22, 24], provide compelling evidence for the superiority of the Volt-Time Disruptive Model for accurate flashover prediction, particularly in complex environments. The remainder of this chapter will build upon this foundation, demonstrating how this advanced modeling approach, combined with detailed simulations and statistical analysis, can be used to improve the design and protection of distribution systems against indirect lightning strikes. The limitations of the simpler 1.5CFO will continue to be highlighted, while the benefits of the volt-time approach are emphasized.

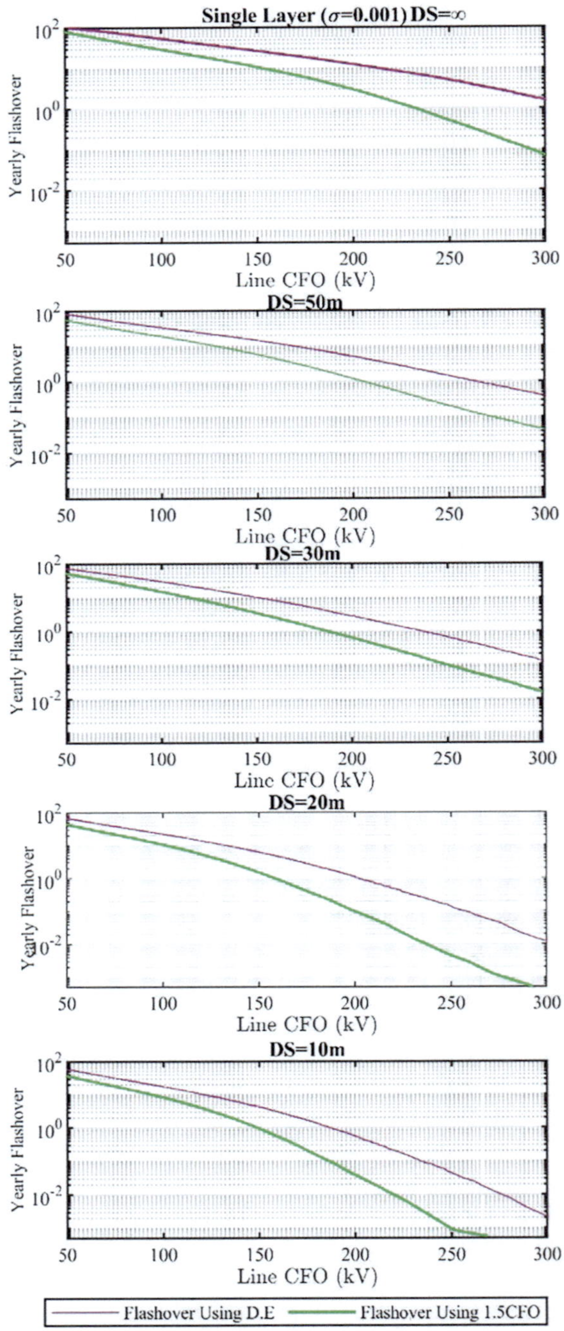

Fig. 6.1 Comparison of annual flashover rates predicted by the 1.5CFO criterion and the Volt-Time Disruptive Model for an overhead distribution line located above a mixed ocean-land propagation path. The distance from the coastline (DS) is varied

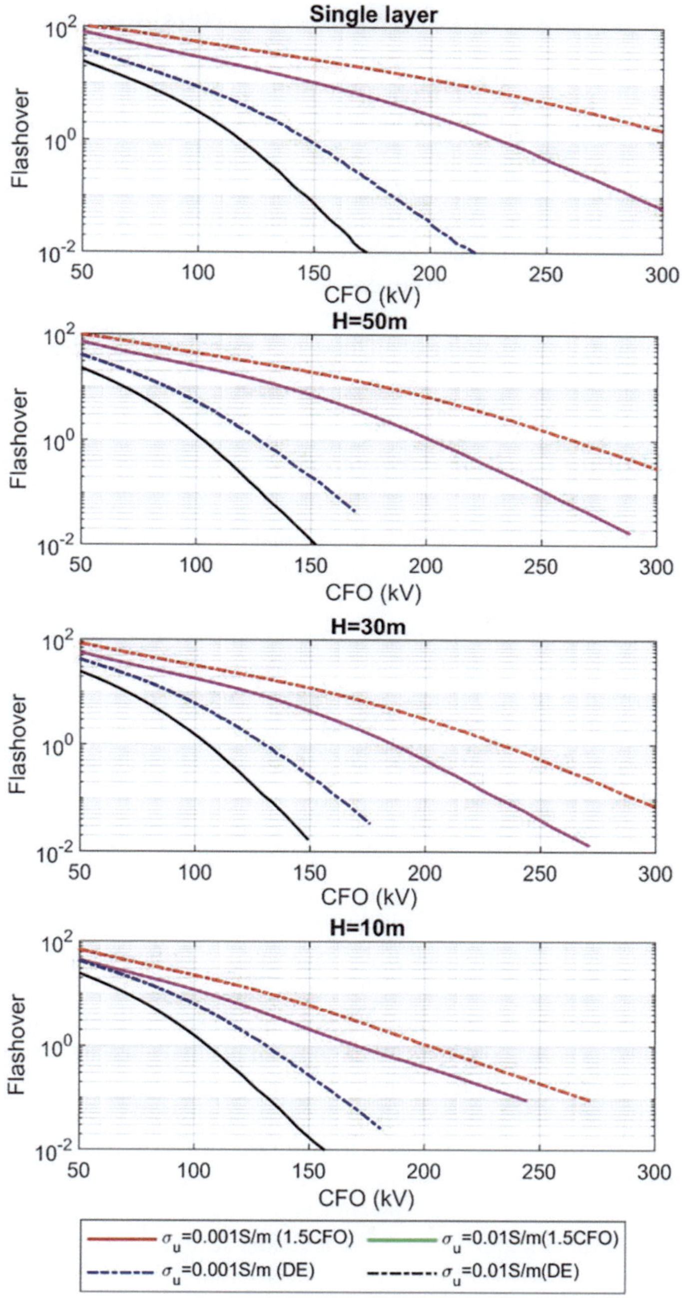

Fig. 6.2 Comparison of annual flashover rates predicted by the 1.5CFO criterion and the Volt-Time Disruptive Model for an overhead distribution line located above horizontally stratified soil. The depth of the upper soil layer (H) is varied, and two different conductivity scenarios are considered

6.3 Key Factors Influencing Flashover Probability

As established in Sect. 6.2, the probability of an indirect lightning strike causing a flashover on an overhead distribution line depends both on the characteristics of the induced voltage and the ability of the line's insulation to withstand that voltage. While Sect. 6.2 focused on the insulator's response (the volt-time characteristic), this section examines the factors that determine the magnitude, waveshape, and duration of the induced voltage itself. These factors can be broadly categorized as (1) lightning current characteristics, (2) ground conditions, and (3) overhead line characteristics. Accurate flashover prediction, and therefore effective lightning protection design, requires a thorough understanding of all these factors and their complex interactions.

6.3.1 Lightning Current Parameters

The lightning return stroke current is the primary source of the electromagnetic fields that induce overvoltage on nearby overhead lines. As discussed extensively in Chap. 1, the return stroke is the most energetic phase of the lightning discharge, characterized by a rapid flow of current from ground to cloud (for negative cloud-to-ground strikes). The waveform of this current—its shape and time variation—is a crucial determinant of the induced voltage characteristics. Chapter 4 presented a detailed methodology for statistically characterizing and generating realistic lightning current waveforms, using the Heidler function and the Cholesky-based Monte Carlo method.

The key parameters that define the return stroke current waveform, and thus significantly influence the induced voltage, are peak current (I_{p}), front time (T_{f}), maximum steepness (S_{m}), and time-to-half value (T_{h}).

These four parameters are not independent; they are statistically correlated. For instance, higher peak currents tend to be associated with faster rise times (shorter front times) and higher maximum steepness.

Chapter 4 presented a detailed methodology for generating statistically representative sets of these correlated parameters using a multivariate log-normal distribution and the Cholesky decomposition technique. This ensures that the simulations used for flashover prediction accurately reflect the natural variability and interdependencies of lightning current characteristics. The use of correlated parameters is crucial for obtaining realistic results; assuming independence between these parameters can lead to inaccurate flashover predictions. In Chap. 4, the typical values and statistical characteristics of these parameters for negative first strokes are summarized.

It is crucial to remember that these are statistical parameters, representing averages and ranges of variability. Individual lightning strokes can deviate significantly from these typical values. This inherent randomness is a primary reason why Monte Carlo simulations, as described in Sect. 6.4, are essential for accurate flashover risk assessment. The Monte Carlo approach allows us to consider a wide range of possible lightning current parameter combinations, reflecting the natural variability of the phenomenon.

6.3.2 *Ground Conditions—A Dominant Influence*

The electrical properties of the ground play a crucially important role in determining the magnitude, waveshape, and propagation characteristics of lightning-induced voltages on overhead distribution lines. The ground acts as a return path for the lightning current and significantly influences the electromagnetic fields generated by the return stroke. This section will delve into the effects of soil resistivity, ground stratification, and the presence of an ocean-land interface—factors that are often oversimplified or neglected in traditional lightning protection studies, but which are essential for accurate flashover prediction, as demonstrated by the research presented throughout this book.

6.3.2.1 Soil Resistivity: A Fundamental Parameter

Soil resistivity (ρ), measured in ohm-meters (Ω m), is a fundamental property that quantifies how strongly the soil *resists* the flow of electric current. It is the inverse of conductivity (σ), where $\sigma = 1/\rho$ (measured in Siemens per meter, S/m. Soil resistivity is not a constant; it varies dramatically, often by several orders of magnitude, depending on several factors:

(a) **Soil Composition**: Different soil types (clay, sand, gravel, rock, loam, etc.) have inherently different resistivities. Clay soils, with their fine particles and ability to retain moisture, generally have lower resistivities (higher conductivities) than sandy, gravelly, or rocky soils.
(b) **Moisture Content**: Water significantly increases the conductivity of soil (lowers its resistivity). Dry soil can have resistivities orders of magnitude higher than saturated soil. This is a critical factor, as soil moisture content can vary significantly with weather conditions (rainfall, drought), seasonal changes, and depth below the surface.
(c) **Temperature**: Soil resistivity generally decreases with increasing temperature (up to a point). This is because increased temperature enhances the mobility of ions within the soil water, facilitating current flow. However, the effect of temperature is usually less pronounced than the effects of moisture content and soil composition.
(d) **Presence of Salts and Minerals**: Dissolved salts and minerals in the soil water can dramatically decrease resistivity by increasing the concentration of charge carriers (ions). This is why seawater is so conductive.

The impact of soil resistivity on lightning-induced voltages is twofold, and both effects are significant:

(1) **Ground Potential Rise (GPR)**: As explained in Chap. 1 and illustrated in Fig. 6.3, when lightning strikes the ground, the current disperses radially outwards from the strike point. The soil's resistance to this current flow creates a voltage gradient, resulting in a ground potential rise (GPR). The magnitude of the GPR is directly proportional to the soil resistivity. Higher resistivity leads to a steeper voltage gradient and a higher GPR for a given lightning current. This elevated ground potential can directly contribute to overvoltage on nearby overhead lines, particularly if the grounding system is inadequate (see Sect. 6.6.3).

(2) **Electromagnetic Field Attenuation**: The soil's conductivity (inverse of resistivity) also affects how electromagnetic fields propagate through the ground and couple with overhead lines. Higher conductivity (lower resistivity) leads to greater attenuation of the electromagnetic fields. This means that the fields decay more rapidly with distance from the lightning channel, reducing the induced voltages on the line. Conversely, lower conductivity (higher resistivity) results in less attenuation, allowing the fields to propagate further with greater strength and induce higher voltages. This effect is significant for the horizontal component of the electric field, which is a major contributor to induced voltages on

Fig. 6.3 Illustration of Ground Potential Rise (GPR) resulting from a lightning strike to ground. The equipotential lines represent areas of equal voltage. Higher soil resistivity (lower conductivity) leads to a steeper voltage gradient and a higher GPR

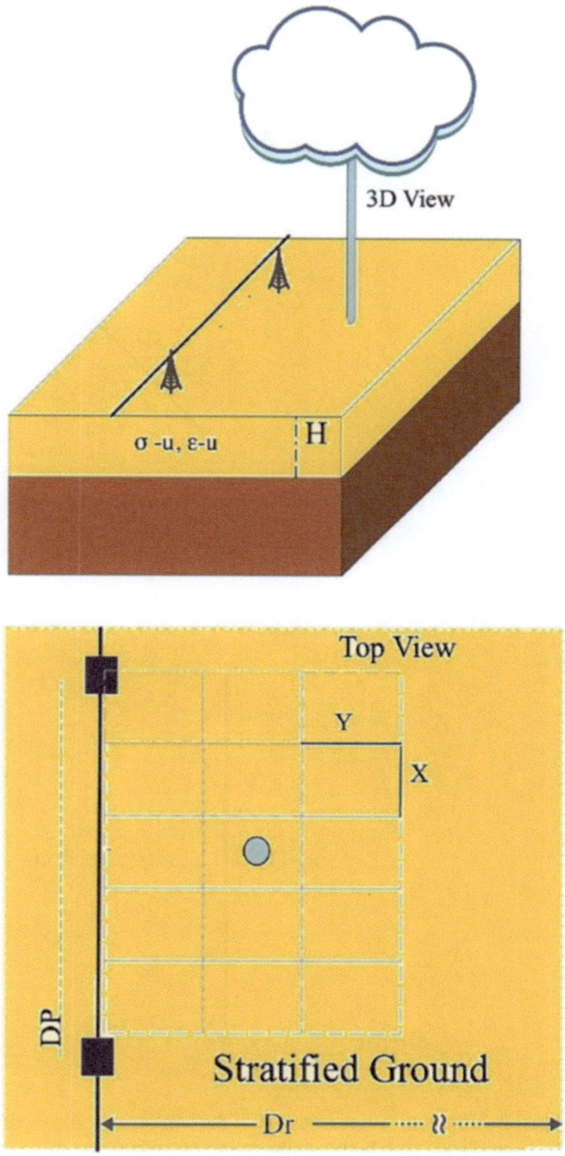

Fig. 6.4 Schematic representation of the two-layer horizontal soil model. The upper layer has a depth (H) and conductivity (σ_u = σ-u) and pemittivity (ε_u = ε-u), while the lower layer extends to infinity and has conductivity (σ_d)

overhead lines. Table 6.1 shows typical soil resistivity values for different soil types [25].

(3) It is crucial to emphasize that these values are only typical ranges. The actual soil resistivity at a specific location can vary significantly and should ideally be measured directly using appropriate techniques, such as the Wenner four-electrode method. Failing to account for the actual soil resistivity can lead to

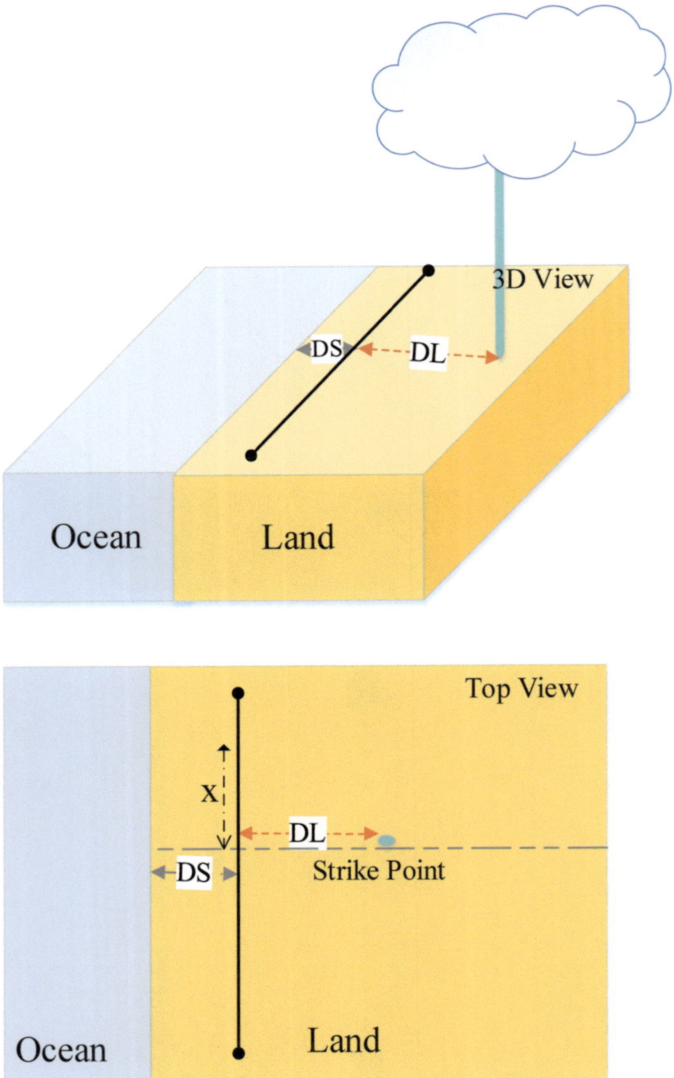

Fig. 6.5 Schematic representation of the two-layer vertical ground model, representing the ocean-land interface. The high conductivity of seawater (σ_{ocean}) contrasts sharply with the lower conductivity (higher resistivity) of typical soil on land (σ_{land})

significant errors in induced voltage calculations and, consequently, in flashover risk assessment.

(4) Assuming a lower resistivity than the actual value will underestimate the induced voltages and the flashover risk, potentially leading to inadequate protection.

Table 6.1 Typical soil resistivity values for different soil types

Soil type	Resistivity range (Ω m)	Description
Clay	15–150	Composed of excellent particles that pack together tightly. It has good water-holding capacity but can be sticky and hard to dig
Loam	15–100	A mixture of sand, silt, and clay, providing a balanced texture. It offers good drainage while retaining moisture and nutrients, making it ideal for plant growth
Sandy clay	50–300	A mixture where sand is the predominant particle size, but with enough clay to provide some cohesion and water retention
Sand	200–3000	Composed of larger, coarse particles with large spaces between them, resulting in excellent drainage but poor water and nutrient retention. It is gritty and light-colored
Gravel and sand	500–5000	Contains a significant amount of gravel (larger rock fragments) mixed with sand. It has very good drainage but very low water retention
Solid rock	10,000+	Dense, consolidated rock material. Resistivity is very high due to the lack of moisture and mobile ions
Wet organic soil	~ 10	Soil with a high content of decomposed plant and animal matter and a high moisture level. The presence of electrolytes in the water contributes to low resistivity
Moist soil	~ 100	Soil with a moderate level of moisture content. Resistivity is higher than that of wet soil due to fewer mobile ions
Dry soil	~ 1000	Soil with very low moisture content. The lack of water significantly reduces the number of mobile ions, leading to high resistivity
Bedrock	~ 10,000+	The solid rock layer underlying the soil. Similar to solid rock, it exhibits high resistivity
Bentonite	2–10	A type of clay known for its high swelling capacity and water absorption. Its unique mineral composition can lead to very low resistivity, especially when wet

6.3.2.2 Ground Stratification: Beyond the Homogeneous Assumption

The assumption of a homogeneous soil (uniform resistivity with depth) is a common simplification in many lightning protection studies, often made for analytical tractability. However, in many real-world scenarios, the soil is stratified, meaning it consists of layers with different electrical properties. This stratification can be due to:

- **Variations in Soil Composition**: Different soil types (e.g., a layer of clay over a layer of sand, topsoil over bedrock) can have significantly different resistivities.
- **Moisture Content Gradients**: The moisture content of soil typically decreases with depth, leading to increasing resistivity with depth, especially after rain.

- **Presence of Subsurface Structures**: Buried pipes, cables, or foundations can alter the local electrical properties.
- **Freezing and Thawing**: In the cold regions freezing can change the soil parameters.

This chapter, and the research presented in [5, 16], focuses on the two-layer horizontal soil model, as illustrated in Fig. 6.4. This model provides a more realistic representation of many common soil profiles than a homogeneous model, while still being relatively tractable for analysis.

The two-layer model is characterized by:

- **Upper Layer**: A layer of depth H (in meters) with a conductivity σ_u (in Siemens per meter, S/m) or resistivity $\rho_u = 1/\sigma_u$.
- **Lower Layer**: A layer extending to infinity (or to a very large depth compared to H), with a conductivity σ_d (in S/m) or resistivity $\rho_d = 1/\sigma_d$.

The presence of these layers, with differing conductivities, introduces significant complexities into the propagation of electromagnetic fields and the calculation of induced voltages. The key effects include:

- **Reflection and Refraction**: Electromagnetic waves generated by the lightning return stroke are reflected and refracted at the interface between the two layers, due to the discontinuity in conductivity. This is analogous to the reflection and refraction of light at the boundary between two different media (e.g., air and water). This creates a complex interference pattern, which can either increase or decrease the induced voltage on the overhead line, depending on the specific parameters (H, σ_u, σ_d, lightning current characteristics, and line location). The reflections can be constructive (adding to the direct wave) or destructive (canceling out part of the direct wave), leading to localized areas of higher or lower field strength. The amount of reflection and refraction depends on the contrast in conductivity between the two layers; a larger contrast leads to stronger reflections.
- **Differential Attenuation**: The electromagnetic fields will be attenuated differently in each layer, depending on their respective conductivities. If the upper layer is more conductive (lower resistivity), it can act as a partial shield, reducing the induced voltages on the line by attenuating the fields before they reach the lower layer. This is because more of the energy is dissipated in the conductive upper layer. Conversely, if the upper layer is less conductive, the fields can penetrate deeper into the ground and potentially induce higher voltages, particularly if the lower layer is more conductive. This is because the less conductive upper layer allows the fields to propagate with less attenuation, and the more conductive lower layer can then "trap" the fields, leading to higher GPR and stronger coupling.
- **Waveshape Distortion**: The reflections and refractions at the layer interface can also distort the waveshape of the induced voltage, altering its rise time, time-to-half value, and overall duration. This is particularly important when considering the volt-time characteristic of insulators, as discussed in Sect. 6.2. A change in waveshape can significantly affect the disruptive effect (DE) and thus the flashover

probability. The reflections can create multiple peaks or oscillations in the voltage waveform.

The research presented in [5, 16], and summarized in Sect. 6.5, demonstrates that both the depth (H) and the conductivity (σ_u) of the upper layer have a *substantial* impact on the annual flashover rate. Different combinations of upper- and lower-layer conductivities (σ_u and σ_d) can lead to significantly different flashover probabilities. For example:

- $\sigma_u > \sigma_d$ **(More Conductive Upper Layer)**: This scenario can provide some shielding, particularly for shallower upper layer depths. However, the shielding effect diminishes as H increases, and for certain combinations of parameters, the flashover rate can actually be higher than for a homogeneous soil with the lower conductivity.
- $\sigma_u < \sigma_d$ **(Less Conductive Upper Layer)**: This scenario can lead to higher induced voltages and flashover rates, particularly for certain values of H, due to the reflections and reduced attenuation in the upper layer. The less conductive upper layer allows the fields to propagate with less attenuation, and the more conductive lower layer can then "trap" the fields, increasing GPR. The analytical formula presented in Sect. 6.5.4 provides a computationally efficient way to estimate flashover rates for different two-layer soil configurations, capturing these complex effects. The results show a complex, non-monotonic relationship between H, σ_u, σ_d, and the flashover rate, highlighting the inadequacy of simplified models that assume a homogeneous soil.

6.3.2.3 Two-Layer Vertical Ground: The Ocean-Land Interface

Coastal regions, and increasingly offshore wind farm installations, present a particularly challenging and important scenario for lightning protection due to the mixed ocean-land propagation path (Fig. 6.5). Seawater possesses a very high conductivity (typically around 4 S/m, or a resistivity of about 0.25 Ω m) compared to typical soil conductivities on land (which often range from 0.01 to 0.001 S/m, or resistivities of 100 to 1000 Ω m). This *drastic* difference in conductivity (often by a factor of 1000 or more) creates a significant discontinuity in the ground's electrical properties, profoundly affecting the propagation of lightning electromagnetic fields and, consequently, the induced voltages on overhead lines. This is a two-layer vertical ground model, and it is a key focus of the research presented in [18, 26].

The key effects of the ocean-land interface include

- **Reduced Attenuation Over Seawater**: The most significant effect is the drastically reduced attenuation of electromagnetic waves propagating over the highly conductive seawater. Unlike over land, where the fields decay relatively quickly due to energy dissipation in the lossy soil, the fields can travel much further over seawater with minimal loss of strength.

  This means that lightning strikes occurring over the ocean, even at considerable distances from the coastline, can induce significantly higher voltages on nearby overhead lines compared to strikes occurring at the same distance over land. The low attenuation effectively extends the "reach" of the lightning strike, increasing the zone of influence.

- **Reflection and Refraction at the Coastline**: he abrupt change in conductivity at the coastline (the land-sea boundary) acts as a boundary for the electromagnetic waves, analogous to the interface between two layers in stratified soil, but with a much stronger contrast. This boundary causes both reflection and refraction of the incident waves:
  - **Reflection**: A portion of the electromagnetic wave energy is reflected back toward the ocean. The amount of reflection depends on the angle of incidence and the conductivity contrast. Because seawater is so conductive, the reflection coefficient is very high, meaning a significant portion of the energy is reflected.
  - **Refraction**: The remaining portion of the wave energy is refracted (bent) as it enters the land. The angle of refraction depends on the conductivity contrast and the angle of incidence.

This reflection and refraction create a complex interference pattern near the coastline. The direct wave from the lightning channel, the reflected wave from the coastline, and the refracted wave traveling through the land can combine constructively (adding together to increase the field strength) or destructively (canceling each other out to some extent), leading to localized areas of enhanced or reduced field strength and, consequently, induced voltage. The precise pattern is highly sensitive to the distance from the coastline, the angle of incidence of the waves (related to the lightning strike location and channel orientation), and the lightning current parameters.

- **Modified Ground Potential Rise (GPR)**: The high conductivity of seawater also significantly alters the distribution of ground currents and the resulting GPR. The current tends to flow preferentially through the seawater, leading to a different GPR profile compared to a homogeneous land surface. This altered GPR can further contribute to the induced voltages on nearby overhead lines in a complex, frequency-dependent manner.
- **Waveshape Distortion**: The reflections at the coastline, combined with the different propagation characteristics over land and water (different attenuation rates and velocities), can significantly distort the waveshape of the induced voltage. This is crucially important because, as emphasized throughout this chapter, the waveshape, not just the peak voltage, is critical in determining flashover probability when using the volt-time model (Sect. 6.2). The ocean-land interface can alter the rise time, fall time, and overall duration of the induced voltage pulse. It can introduce multiple peaks or oscillations due to reflections.

These changes in waveshape can dramatically affect the disruptive effect (DE) calculated in the volt-time model.

- **Distance from the Coastline (DS)**: The distance from the coastline (DS), defined as the horizontal distance between the overhead line and the ocean-land boundary,

is a critical parameter in determining the magnitude and waveshape of induced voltages in coastal areas. As the line moves closer to the coast (smaller DS), the influence of the seawater becomes more pronounced. However, the overall effect on the annual flashover rate is complex and non-monotonic. It doesn't simply increase or decrease linearly with DS. This is because of the interplay between the several competing factors: reduced attenuation over water (which tends to increase voltages), reflection/refraction at the coastline (which can cause both increases and decreases), and the altered GPR distribution. The research presented in [18, 26] demonstrates this complex relationship, showing that the annual flashover rate can vary significantly with DS, and the optimal placement of protective devices (surge arresters) is highly dependent on this distance.
- **Invalidation of Simplifying Assumptions**: It is essential to recognize that the presence of the ocean-land interface invalidates many of the simplifying assumptions commonly used in traditional lightning protection studies. Assumptions of a homogeneous, perfectly conducting, or even finitely conducting ground with uniform properties are completely inadequate in this scenario. Accurate flashover prediction in coastal regions requires the use of advanced modeling techniques, such as 3D FEM (Sect. 6.4), that can properly account for the mixed propagation path (high-conductivity seawater and lower-conductivity land) and the complex interactions between the lightning electromagnetic fields and this discontinuous ground. The analytical formula presented in Sect. 6.5.3 is specifically designed to address this complex scenario, providing a computationally efficient yet accurate tool for engineers.

6.3.3 Overhead Line Characteristics

While ground conditions and lightning current parameters are the dominant factors influencing induced voltages from indirect lightning strikes, certain characteristics of the overhead distribution line itself also play a role, albeit often a secondary one.

6.3.3.1 Line Height and Conductor Size

The height of the conductors above ground and their diameter has a relatively minor influence on the induced voltages resulting from indirect lightning strikes, compared to the strong effects of ground conditions and lightning current parameters.

This is because the induced voltages are primarily driven by the fields generated by the lightning channel and the ground currents, rather than direct interception of the channel (which is the primary concern for direct strikes).

Higher lines are, geometrically, exposed to a slightly larger portion of the incident electromagnetic field. Similarly, larger diameter conductors present a slightly larger cross-sectional area for coupling with the field. These factors can, in principle, lead to marginally higher induced voltages. However, for the typical range of line heights (e.g., 10–20 m) and conductor sizes used in distribution systems, this effect is often secondary compared to the influence of ground resistivity, stratification, and the presence of an ocean-land interface. In many simulation studies focusing on indirect strikes, a constant, representative line height and conductor size are assumed to simplify the model without significantly compromising accuracy, provided that the dominant effects of ground conditions are accurately modeled.

6.3.3.2 Pole Spacing (DP)

The distance between supporting poles (DP) does not directly affect the magnitude of the induced voltage at a specific point on the line. The induced voltage at a given location is primarily determined by the electromagnetic fields at that point, which are governed by the lightning current, strike location, and ground conditions.

However, pole spacing has a significant indirect influence on the overall flashover probability of the entire line. Closer pole spacing (smaller DP) means there are more insulators along a given length of line. Each insulator represents a potential flashover point. Therefore, even if the induced voltage at each individual insulator is slightly lower due to other factors (e.g., a more conductive ground), the overall probability of at least one flashover occurring along the line can be higher with closer pole spacing, simply because there are more opportunities for a flashover to occur.

This is a crucial point, often overlooked in simplified analyses. It highlights the difference between considering the voltage at a single point versus the overall system reliability. The research presented in [5, 16, 18, 26] demonstrates this effect, showing that decreasing pole spacing can increase the annual flashover rate, even when other factors might suggest lower individual induced voltages. This is a probabilistic effect: more insulators increase the likelihood of at least one exceeding it withstand capability.

6.3.3.3 Insulator Characteristics

Overhead distribution lines utilize insulators, typically made of porcelain, glass, or polymer materials, to electrically isolate the energized conductors from the grounded support structures (poles or towers).

The critical flashover voltage (CFO) of an insulator is a key parameter, representing the voltage level at which the insulator is likely to experience a flashover (electrical breakdown). As discussed in Sect. 6.2, the traditional 1.5CFO criterion provides a simplified estimate of the flashover voltage, but the more accurate volt-time disruptive model considers the dynamic behavior of the insulator and the shape of the induced voltage waveform. The choice of insulator type and its CFO rating is a critical design consideration for ensuring lightning resilience. Higher CFO insulators offer greater protection, but at a higher cost.

The main insulator types are

- **Porcelain Insulators**:
  - *Strengths*: Good mechanical strength, relatively inexpensive, long history of use, and well-understood performance.
  - *Weaknesses*: Susceptible to surface contamination (pollution, salt spray), which can significantly reduce their flashover voltage. Lower impulse flashover voltage compared to some newer materials. Brittle and prone to cracking or shattering under mechanical stress.
- **Glass Insulators**:
  - *Strengths*: Similar characteristics to porcelain. Slightly better resistance to contamination in some cases. It can be visually inspected for internal cracks.
  - *Weaknesses*: Similar to porcelain.
- **Polymer (Composite) Insulators:**
  - *Strengths: Higher Hydrophobicity*: Polymer insulators have a hydrophobic surface, meaning they repel water. This prevents the formation of continuous water films on the insulator surface, which can significantly reduce the flashover voltage, especially in polluted environments. Lighter Weight: Reduce mechanical load on structures, allowing for potentially longer spans or lighter towers. Higher Mechanical Strength: Some polymer insulators offer higher mechanical strength and resistance to vandalism or impact damage. Better performance under pollution.
  - *Weaknesses:* Long-term aging and degradation mechanisms are still under investigation, particularly for newer formulations. Can be more susceptible to certain types of environmental degradation (e.g., UV radiation, chemical attack) if not properly formulated and protected. May be more expensive initially, though life-cycle costs can be lower due to reduced maintenance and replacement.

The selection of an appropriate insulator involves considering the CFO, the volt-time characteristic, the expected environmental conditions (pollution, humidity, salt spray), mechanical requirements, and cost. The advanced modeling techniques presented in this chapter allow for a more informed choice by providing accurate predictions of induced voltage waveforms, enabling the use of the volt-time model for flashover assessment.

6.3.3.4 Presence of Shield Wires and Surge Arresters

While the primary function of shield wires (also known as overhead ground wires) is to intercept direct lightning strikes and prevent them from reaching the phase conductors (as discussed in Chap. 2), they also provide a degree of protection against indirect strikes by reducing the induced voltages on the phase conductors. Shield wires are conductors installed above the phase conductors and connected to ground at regular intervals (typically at each pole).

- **Mechanism**: Shield wires reduce induced voltages through:
  - **Electrostatic Shielding**: The grounded shield wire alters the electric field distribution, reducing capacitive coupling.
  - **Electromagnetic Shielding**: The shield wire provides a low-impedance path to ground for induced currents, reducing inductive coupling.
- **Effectiveness**: The effectiveness depends on shield wire height, position, number of shield wires, and grounding resistance.

  Surge arresters (or surge protective devices, SPDs), as detailed in Chap. 2, are essential for limiting overvoltages. They are typically metal-oxide varistors (MOVs), most commonly made of zinc oxide (ZnO), that provide a low-impedance path to ground when the voltage exceeds their clamping voltage.
- **Coordination**: Shield wires and surge arresters work together. Shield wires reduce the frequency and magnitude of overvoltage, while surge arresters limit the voltage that reaches the equipment. Proper coordination between them is crucial for adequate protection.

6.3.3.5 Summary of Factors Influencing Flashover Probability

In summary, predicting lightning-induced flashover on overhead distribution lines requires a comprehensive understanding of the interplay between multiple factors:

- **Lightning Current Parameters**: Peak current (I_p), front time (T_f), maximum steepness (S_m), and time-to-half value (T_h) are the primary drivers of the induced voltage magnitude and waveshape. These parameters are statistically correlated and exhibit significant natural variability.
- **Ground Conditions**: Soil resistivity, ground stratification (horizontal layers), and the presence of an ocean-land interface significantly influence the propagation of electromagnetic fields and the resulting induced voltages. These factors often dominate the overall response and must be accurately modeled.
- **Overhead Line Characteristics**: Line height and conductor size have a relatively minor influence on indirect strikes. Pole spacing affects the overall flashover probability of the line, even if it doesn't directly change the induced voltage at a single point. Insulator characteristics (CFO, volt-time curve) are fundamental to determining flashover, and the presence and design of shield wires and surge arresters are crucial mitigation measures.

The complexity of these interacting factors necessitates the use of advanced modeling techniques, such as those described in Sect. 6.4, to accurately predict flashover risk and to design effective lightning protection systems. Simplified approaches, like the 1.5CFO criterion, are often inadequate, particularly in non-uniform ground conditions. The analytical formulas presented in Sect. 6.5 provide a valuable tool for bridging the gap between complex simulations and practical engineering applications.

6.4 Advanced Modeling Techniques for Induced Voltage Calculation

Accurate prediction of lightning-induced voltages on overhead distribution lines, particularly in the presence of complex ground conditions, requires sophisticated modeling techniques that can capture the intricate electromagnetic interactions involved. Simplified analytical methods, as reviewed in Chap. 3, are insufficient to represent these complexities accurately. This research utilizes a combination of advanced numerical modeling techniques to overcome these limitations:

- **3D Finite Element Method (FEM)**: For the full-wave solution of Maxwell's equations in complex geometries.
- **Modified Transmission Line Model with Exponential Decay (MTLE)**: For representing the lightning return stroke channel current.
- **Heidler Function**: For modeling the channel-base current waveform.
- **Monte Carlo Simulation**: For incorporating the statistical variability of lightning parameters and strike locations.

This section details these modeling techniques and their implementation within the framework of this research, drawing heavily on the methodologies described in [16, 18, 24, 26].

6.4.1 Overview of Modeling Approaches

Chapter 3 provided a comprehensive review of various methods for calculating lightning-induced voltages, categorizing them as:

- **Analytical Methods:** These methods, such as the Bergeron method and the traveling wave method (mentioned in Chap. 3), rely on simplified representations of the lightning channel, overhead line, and ground. They often assume a perfectly conducting ground and employ analytical expressions derived from Maxwell's equations under these simplifying assumptions. While computationally efficient (requiring minimal computational resources and time), analytical methods have limited accuracy, particularly when dealing with complex geometries, non-uniform ground conditions (stratification, ocean-land interfaces), and non-standard lightning current waveforms. They are useful for initial estimations but are not suitable for detailed, accurate analysis.
- **Numerical Methods**: These methods provide a more accurate and flexible approach by solving Maxwell's equations numerically, without the need for the restrictive simplifying assumptions of analytical methods. Common numerical techniques used in lightning studies include
  - **Finite-Difference Time-Domain (FDTD):** A time-domain method that discretizes both space and time and solves Maxwell's equations directly using

finite-difference approximations. FDTD is well-suited for modeling transient phenomena like lightning and can handle complex geometries and material properties [27–33].

- **Method of Moments (MoM)**: A frequency-domain method that solves integral equations for the currents and charges on conductors. MoM is particularly efficient for modeling wire antennas and transmission lines, but it can become computationally expensive for complex geometries [4, 34, 35].
- **Finite Element Method (FEM)**: A versatile numerical method that divides the problem domain into small elements (finite elements) and approximates the solution within each element using simple functions. FEM is particularly well-suited for handling complex geometries, inhomogeneous materials (like stratified soil or the ocean-land interface), and a wide range of boundary conditions.

The research presented in this chapter primarily utilizes the **3D Finite Element Method (FEM)**, due to its ability to accurately model the complex scenarios under consideration.

6.4.2 The 3D Finite Element Method (FEM)

The Finite Element Method (FEM) is a powerful numerical technique for solving partial differential equations (PDEs), such as Maxwell's equations, which govern electromagnetic phenomena.

Unlike analytical methods that seek exact, closed-form solutions (which are often impossible to obtain for complex problems), FEM provides an approximate solution by dividing the problem domain into smaller, simpler subdomains.

The core idea of FEM is to discretize the problem domain (the region of space where we want to calculate the electromagnetic fields—in this case, the air, ground, and lightning channel) into a large number of small, interconnected *finite elements.* These elements can be triangles or quadrilaterals in 2D, or tetrahedra, hexahedra, prisms, or pyramids in 3D. The collection of these elements forms a mesh. The choice of element type and mesh density depends on the geometry of the problem and the desired accuracy.

Within each element, the unknown fields (e.g., the electric and magnetic field vectors) are approximated using simple functions, typically polynomials. These are called basis functions or shape functions. The choice of basic functions affects the accuracy and computational efficiency of the method.

Maxwell's equations, which are originally in differential or integral form, are then transformed into a set of algebraic equations by applying a weighted residual method (e.g., Galerkin's method) within each element. This involves multiplying the governing equations by the basics functions and integrating over the element volume. This process effectively converts the continuous problem (infinite degrees of freedom) into a discrete problem (finite degrees of freedom).

The resulting element equations are then assembled into a global system of equations, representing the entire problem domain. This global system is a large, sparse matrix equation (sparse because each element only interacts with its immediate neighbors). This system of equations is then solved using numerical techniques, such as direct solvers (e.g., Gaussian elimination) or iterative solvers (e.g., conjugate gradient method), to obtain the approximate values of the fields at the nodes of the mesh. The advantages of using FEM for lightning-induced voltage calculations include

- **Complex Geometries**: FEM can readily handle complex geometries, such as
  - Irregular coastlines (ocean-land interface).
  - Uneven terrain (hills, valleys)—although this is not the primary focus of *this* chapter, the framework allows for its inclusion.
  - The presence of nearby structures (buildings, towers)—again, not explicitly modeled in the core simulations presented here, but the FEM framework can incorporate them.
  - Horizontally stratified soil with varying layer depths and conductivities.
- **Inhomogeneous Materials**: FEM can easily incorporate variations in material properties, such as
  - The different conductivities of soil layers in a stratified ground model.
  - The drastic conductivity difference between seawater and land.
  - The (potentially) frequency-dependent behavior of soil properties (although the primary focus here is on constant conductivities for each layer).
- **Full-Wave Solution:** FEM provides a full-wave solution of Maxwell's equations, meaning it accounts for all relevant electromagnetic phenomena, including:
  - Radiation of electromagnetic waves from the lightning channel.
  - Induction (both electric and magnetic coupling) between the lightning channel, the ground, and the overhead line.
  - Wave propagation and attenuation in lossy media (the ground).
  - Reflection and refraction of waves at material interfaces (e.g., between soil layers or at the coastline).
- **Flexibility**: FEM can be adapted to a wide range of boundary conditions and excitation sources, making it suitable for modeling various lightning scenarios and different return stroke models.

  The specific FEM implementation used in this research is the **COMSOL Multiphysics** software package [36], a commercially available, widely used, and validated tool for Multiphysics simulations. Within COMSOL, the **RF Module** is employed, specifically the **Time Domain Electromagnetic Wave (TEMW)** physics interface. This interface is designed for solving the time-dependent Maxwell's equations in the presence of electromagnetic waves, making it ideally suited for modeling the transient electromagnetic fields generated by lightning.

Key aspects of the COMSOL implementation include

- **Governing Equations**: The TEMW interface solves the time-dependent Maxwell's equations in their vector potential formulation. This formulation is mathematically equivalent to the standard electric and magnetic field formulation but is often more convenient for numerical implementation, particularly in the presence of conductive media.
- **Boundary Conditions**: Scattering boundary conditions are applied to the outer boundaries of the simulation domain. These conditions simulate an open boundary, allowing electromagnetic waves to propagate outwards without reflections.

This is crucial for avoiding spurious reflections from the artificial boundaries of the model, which would contaminate the results and lead to inaccurate induced voltage calculations. The scattering boundary conditions effectively mimic an infinitely large domain, preventing artificial reflections.

- **Mesh Generation**: COMSOL automatically generates a mesh of tetrahedral elements. The mesh density (the size of the elements) is a critical parameter for accuracy. A finer mesh (smaller elements) generally leads to more accurate results, as the fields are better approximated within each element. However, a finer mesh also significantly increases the computational cost (memory requirements and solution time). The research papers [16, 18, 24, 26] describe mesh optimization techniques used to ensure sufficient accuracy while maintaining reasonable computation times. These techniques typically involve adaptive mesh refinement, where the mesh is automatically refined in regions where the fields are changing rapidly (e.g., near the lightning channel and the overhead line) and coarsened in regions where the fields are relatively uniform.
- **Simplified Line Model**: To reduce computational complexity, the overhead distribution line is modeled as a thin wire conductor, rather than resolving its full three-dimensional geometry (e.g., the actual diameter of the conductor). This simplification is justified because the diameter of the conductor is typically much smaller than the wavelengths of interest in the lightning-induced voltage calculations. Furthermore, the detailed modeling of the poles and grounding network is not included in the FEM simulations themselves. Instead, the flashover criteria (1.5CFO and Volt-Time, as discussed in Sect. 6.2) are applied to the calculated induced voltages at the locations of the insulators to determine whether a flashover occurs. This approach, consistent with the IEEE 1410 guidelines [7], implicitly accounts for the presence of insulators, poles, and the grounding system, significantly reducing the computational burden without sacrificing accuracy in the prediction of flashover occurrence. The FEM model focuses on accurately calculating the induced voltage, and the separate flashover criteria determine whether that voltage will cause a breakdown.

6.4.3 Lightning Channel and Current Modeling

The lightning return stroke channel is represented using the **Modified Transmission Line Model with Exponential Decay (MTLE)**, as discussed in Chap. 3. This is an engineering model that describes the current distribution along the channel as a function of time and height. Engineering models, unlike more complex physical models, provide a simplified, yet reasonably accurate, representation suitable for electromagnetic field calculations. The MTLE model makes the following assumptions:

- A vertical channel: While Chapter 5 explores inclined channels, the core simulations used for deriving the analytical formulas in this chapter (and in [5, 16, 18, 26]) assume a vertical channel. This simplifies the calculations, but is a limitation to be aware of.
- Current *propagation*: A current pulse propagates upwards along the channel with a defined velocity (v).
- *Exponential Decay:* The current decays exponentially with height, characterized by a decay constant (λ). This decay accounts for the loss of charge from the channel due to corona and other processes.

The MTLE model equation is

$$I(z,t) = I_0\left(0, t - \frac{z}{v}\right)e^{\frac{-z}{\lambda}} \tag{6.2}$$

where

- $I(z,t)$: The current at height z along the channel at time t.
- $I(0,t)$: The current at the base of the channel (ground level) at time t. This is the input to the MTLE model.
- v: The return stroke propagation velocity. Typical values range from 1/3 to 2/3 the speed of light (approximately 1×10^8 to 2×10^8 m/s). The research papers often use a value of 1.5×10^8 m/s [5, 16, 18, 26].
- z: The height above ground.
- λ: The decay constant represents the characteristic length over which the current decreases with height. Typical values range from a few hundred meters to a few kilometers (often 2000 m is used).
- $t - z/v$: Represents the time delay for the current wave to propagate from the base of the channel to height z.

The MTLE model, while a simplification, provides a reasonable approximation for calculating the electromagnetic fields, particularly at distances relevant for induced voltage calculations on overhead lines.

The *input* to the MTLE model is the channel-based current, $I(0,t)$. This research utilizes the Heidler function to represent the channel-base current waveform. The Heidler function, as introduced in Chap. 3 and detailed in Chap. 4, is a widely used and accepted model (adopted by IEC 62305 [37]) for representing typical lightning current waveforms. It is defined by the following equation:

$$I(0,t) = \frac{I_0}{\eta_1}\frac{(t/\tau_{11})^{n_1}}{1+(t/\tau_{11})^{n_1}}\exp\left(\frac{-t}{\tau_{12}}\right) \tag{6.3}$$

and

$$\eta_1 = \exp\left[\left(\frac{-\tau_{11}}{\tau_{12}}\right)\left(\frac{n_1 \times \tau_{12}}{\tau_{11}}\right)^{1/(n_1)}\right] \tag{6.4}$$

where

- $I(0, t)$: The channel-base current as a function of time.
- I_0: An amplitude parameter (related to, but not exactly equal to, the peak current I_p).
- τ_{11}: A time constant that primarily controls the front time (rise time) of the current waveform.
- τ_{12}: A time constant that primarily controls the *decay time* (tail time) of the current waveform.
- n_1: An exponent that controls the *steepness* of the current rise. Typical values range from 2 to 10. Higher values of n_1 result in a steeper, faster-rising current.
- η_1: A correction factor that ensures that the peak current of the Heidler function is close to I_0. It is a function of τ_{11}, τ_{12}, and n_1.

The Heidler function is flexible and can represent a wide range of current waveforms by adjusting its parameters. However, determining the appropriate values for these parameters to match specific lightning characteristics (I_p, T_f, S_m, T_h) is not straightforward. The parameters are interrelated, and a change in one parameter affects the others.

Chapter 4 provides a detailed methodology for determining the appropriate Heidler function parameters. This methodology involves

1. **Statistical Analysis of Measured Data**: Using data from lightning measurements (e.g., from instrumented towers), the statistical distributions of the key lightning current parameters (I_p, T_f, S_m, T_h) are determined. These distributions are typically found to be log-normal, and the correlations between the parameters are also quantified.
2. **Cholesky Decomposition**: The Cholesky decomposition technique is used to generate correlated random variables that follow the specified multivariate log-normal distribution. This ensures that the generated sets of parameters (I_p, T_f, S_m, T_h) reflect the natural relationships between these parameters (e.g., higher peak currents tend to have faster rise times).
3. **Genetic Algorithm (GA)**: For each set of generated lightning current characteristics (I_p, T_f, S_m, T_h), a Genetic Algorithm (GA) is used to find the optimal set of Heidler function parameters (I_0, τ_{11}, τ_{12}, n_1) that best reproduce those characteristics. The GA is an optimization algorithm that mimics the process of

natural selection to find the best solution to a problem. In this case, the "best" solution is the set of Heidler parameters that minimizes the difference between the desired lightning current characteristics and the characteristics of the waveform produced by the Heidler function. The objective function used in the GA is described in detail in Chap. 4.

This process ensures that the simulations use realistic and statistically representative lightning current waveforms as input, which is crucial for accurate induced voltage calculations.

6.4.4 Induced Voltage Calculation

As it is described in previous sections in each scenario the induced voltage due to lightning should be calculated. As Chap. 3 described, there is different methods for computation of induced voltage. In this section, we explained the used method.

To efficiently compute the induced voltage on the overhead line for arbitrary lightning current waveforms (represented by the Heidler function with varying parameters), this method utilizes a technique based on the step response of the system and the Fourier transform. Directly solving the 3D FEM model for every different Heidler function waveform would be computationally prohibitive, given the large number of scenarios required for the Monte Carlo simulations. The step-response/Fourier transform method significantly reduces computational cost.

The procedure is as follows:

1. **Step Response Calculation**: The 3D FEM model (described in Sect. 6.4.2) is initially solved for a unit step function input current at the base of the lightning channel. This means the current is instantaneously switched on from 0 to 1 (dimensionless) at time $t = 0$. The resulting time-varying induced voltage on the overhead line, $V(U(t))$, is calculated and stored. This is the step response of the system—the response of the system (the overhead line and its environment) to a unit step input.
2. **Frequency Response Calculation**: The frequency response of the system, $H(\omega)$, is then calculated by taking the Fourier Transform of the step response, $V(U(t))$, and dividing it by the Fourier Transform of the unit step function (which is $1/j\omega$, where j is the imaginary unit and ω is the angular frequency):

$$H(\omega) = \frac{\mathrm{FFT}(V(U(t)))}{\mathrm{FFT}(U(t))} \tag{6.5}$$

   The frequency response, $H(\omega)$, represents how the system (the overhead line, the ground, and the surrounding environment) responds to different frequency components.
3. **Arbitrary Waveform Calculation**: Once the frequency response, $H(\omega)$, is known, the induced voltage for any arbitrary channel-base current waveform, $I(0,$

t) (e.g., a Heidler function with specific parameters), can be efficiently calculated using the following steps:

(a) Calculate the Fourier Transform of the desired channel-base current, FFT[$I(0, t)$]. This transforms the time-domain current waveform into its frequency-domain representation.
(b) Multiply the frequency response, $H(\omega)$, by the Fourier Transform of the current, FFT[$I(0, t)$]. This is a simple element-wise multiplication in the frequency domain.

    This step effectively applies the system's response to each frequency component of the input current.
(c) Take the *inverse* Fourier Transform (IFFT) of the result to obtain the time-domain induced voltage, $V(I(0, t))$:

$$V(I(0,t)) = \text{IFFT}(H(\omega) * \text{FFT}[I(0,t)]) \tag{6.6}$$

This equation represents the core of the efficient calculation method. It states that the induced voltage due to any lightning current waveform can be obtained by multiplying the frequency response of the system (obtained *once* from the step response) by the frequency spectrum of the desired current waveform and then transforming the result back to the time domain.

This approach is significantly more efficient than directly solving the 3D FEM model for every different current waveform. The computationally expensive FEM simulation is performed only once to obtain the step response (and thus the frequency response). The induced voltages for all subsequent scenarios (different lightning current parameters, different strike locations) are then calculated quickly and efficiently using fast Fourier transforms (FFTs) and inverse fast Fourier transforms (IFFTs), which are computationally inexpensive operations. This efficiency is crucial for enabling the Monte Carlo simulations, which require thousands of individual induced voltage calculations.

6.4.5 Monte Carlo Simulation

The inherent randomness of lightning necessitates a statistical approach for accurate flashover risk assessment. Lightning parameters, such as strike location, peak current, rise time, and tail time, vary significantly from one strike to another. To account for this variability, and to obtain statistically meaningful results (e.g., the annual flashover rate, which is a probabilistic quantity), this chapter employs Monte Carlo simulations.

The Monte Carlo method, in this context, involves the following steps:

1. **Scenario Generation**: A large number of lightning scenarios (typically thousands or tens of thousands) are generated. For each scenario, the key parameters are randomly selected from their respective statistical distributions:
   (a) **Lightning Current Parameters**: As detailed in Chap. 4, correlated sets of lightning current parameters (I_p, T_f, S_m, T_h) are generated using the multivariate log-normal distribution and the Cholesky decomposition technique. This ensures that the generated parameters reflect the natural variability and interdependencies observed in real lightning strikes.
   These parameters are then used to define the Heidler function (Eq. 6.3), which represents the channel-base current, $I(0, t)$, for that specific scenario.
   (b) **Strike Location**: The location of the lightning strike is randomly selected within a defined area around the overhead line. A uniform distribution is typically assumed, meaning that each location within the area has an equal probability of being struck. The size of this area is chosen to be large enough to include all strikes that could potentially induce significant voltages on the line (typically extending several hundred meters or even kilometers from the line, depending on the ground conditions and the sensitivity of the analysis).
   (c) **Striking Distance**: For each generated scenario, it is determined whether the strike is direct or indirect. The striking distance concept, as introduced in Chap. 1 and discussed in Chap. 3, is used. The striking distance is the distance from the tip of the descending lightning leader to the object (line or ground) at which the final jump occurs essentially; it defines the zone of attraction of the object. Several empirical models are commonly used to calculate the striking distance, including:
      (i) **Electro-Geometric Model (EGM)**
      (ii) **Leader Progression Model (LPM)**
      (iii) **Eriksson Model**
      (iv) **Rizk Model**.

These models typically relate the striking distance to the peak lightning current (I_p) and the height of the object (h). Table 6.2 summarizes the equations for these models.

If the distance between the lightning channel and the line is less than the calculated striking distance, the strike is considered direct. Direct strikes are typically excluded from the analysis of indirect flashovers, as this chapter focuses on indirect effects.

Table 6.2 Striking distance equations for different models

EGM	$r=\sqrt{r_s^2-(r_g-h)^2}$	$r_s=10I_0^{0.65}$	$r_g=0.9r_s$	
LPM	$r=c+aI_p^b$	$a=0.67h^{0.6}$	1	$c=3h^{0.6}$
Eriksson		$a=0.67h^{0.6}$	0.74	0
Rizk		$a=1.57h^{0.45}$	0.69	0

If the distance is greater than the striking distance, the strike is considered indirect, and the simulation proceeds to the next step.

2. **Induced Voltage Calculation**: For each generated lightning scenario (characterized by a specific set of lightning current parameters and a strike location), the time-varying induced voltage on the overhead line, $U(t)$, is calculated. This is done using the 3D FEM model (described in Sect. 6.4.2) and the efficient step-response/Fourier Transform method (described in Sect. 6.4.4). This step provides the crucial input for the flashover determination.
3. **Flashover Determination**: For each calculated induced voltage waveform, $U(t)$, the flashover criterion is applied. As discussed in detail in Sect. 6.2, two primary criteria are considered:

   (a) **1.5CFO Criterion:** A flashover is assumed to occur if the peak value of $U(t)$ exceeds 1.5 times the insulator's CFO.

   (b) **Volt-Time Disruptive Model:** A flashover is assumed to occur if the disruptive effect (DE), calculated by integrating the voltage waveform according to Eq. 6.2, exceeds the critical value DE*.

   The research presented in this chapter emphasizes the use of the Volt-Time Disruptive Model due to its superior accuracy and physical realism.

4. **Probability Calculation:** Each individual lightning scenario is assigned a probability of occurrence based on:

   (a) **The probability density functions (PDFs) of the lightning current parameters:** The multivariate log-normal distribution, as described in Chap. 4, defines the probability of occurrence of each specific combination of I_{p}, T_{f}, S_{m}, and T_{h}.

   (b) **The probability of the strike location**: Assuming a uniform distribution of strike locations within the defined area, the probability of a strike within a small sub-area (e.g., a rectangle, as used in the simulations) is proportional to the area of that sub-area. This is often represented as

$$P_A = (\text{Area of Rectangle})/(\text{Total Striking Area}) \tag{6.7}$$

   where P_A is the probability of striking.

5. **Annual Flashover Rate Calculation:** The annual flashover rate (F_{p}) is the ultimate output of the Monte Carlo simulation. It is calculated by summing the probabilities of all scenarios that result in a flashover, and multiplying by factors to give [5, 16]:

$$F_P = 100 \times N_{\mathrm{g}} \times \sum_{1}^{N} \sum_{1}^{R} P_A * f(i) \tag{6.8}$$

- N is number of scenarios.

- R is the rectangles considered as the struck point around the line.
- N_g is ground flash density.
- $f(i)$ *is* the probability of a lightning current resulting in a flashover on the insulator string (based on the DE or 1.5 times the CFO criterion).

This provides a statistically robust estimate of the expected number of flashovers per year per 100 km of line length, for the given line configuration, ground conditions, and lightning environment.

The Monte Carlo approach, combined with the detailed 3D FEM modeling and accurate representation of lightning current parameters, provides a powerful and comprehensive methodology for assessing the risk of indirect lightning-induced flashover. The large number of simulations ensures that the inherent randomness of lightning is properly accounted for, leading to statistically reliable results.

6.5 Analytical Formulas for Flashover Rate Estimation

The preceding sections have detailed the advanced modeling techniques used to accurately calculate lightning-induced voltages and flashover probabilities. While the combination of 3D FEM simulations and Monte Carlo analysis (Sect. 6.4) provides a high degree of accuracy, it can be computationally demanding and time-consuming, particularly when evaluating numerous scenarios or performing sensitivity analyses. For practical engineering applications, such as preliminary design, rapid risk assessments, and optimization studies, simplified, yet accurate, analytical formulas are highly desirable. Such formulas allow engineers to quickly estimate flashover rates without the need for extensive numerical simulations.

This section presents two such analytical formulas, derived from the extensive research and simulations documented in [5, 16, 18]. These formulas are specifically tailored for:

1. **Mixed Ocean-Land Environments**: This formula addresses the complex scenario of overhead distribution lines located near coastlines, where the high conductivity of seawater significantly alters electromagnetic field propagation.
2. **Horizontally Stratified Soil**: This formula addresses the common situation where the soil consists of two horizontal layers with different conductivities.

Both formulas are based on the Generalized Extreme Value (GEV) distribution, a statistical distribution well-suited for modeling extreme events like lightning-induced overvoltage.

6.5.1 Motivation: The Need for Simplified Formulas

As discussed in Sect. 6.4, the detailed modeling approach (3D FEM coupled with Monte Carlo simulations) offers high accuracy but has practical limitations:

- **Computational Cost**: FEM simulations, especially when combined with Monte Carlo analysis requiring thousands of runs, demand significant computational resources and time.
- **Complexity**: Setting up and executing FEM simulations requires specialized software (like COMSOL) and expertise in electromagnetic modeling.
- **Parameter Sensitivity Studies**: Investigating the impact of changing parameters (e.g., soil resistivity, distance to coast) requires rerunning the entire simulation set, which is inefficient.
- Analytical formulas, in contrast, offer significant advantages:
- **Computational Efficiency**: Formulas can be evaluated extremely quickly, even with simple calculators or spreadsheet software. This enables rapid assessment of flashover rates under various conditions.
- **Simplicity and Accessibility**: Formulas provide a direct and transparent relationship between input parameters (e.g., CFO, distance from coastline, soil conductivities) and the output (annual flashover rate). This makes them easy to understand and use, even for engineers who are not specialists in electromagnetic modeling.
- **Insight and Sensitivity Analysis**: The analytical form of the formulas facilitates understanding how each parameter influences the flashover rate. This allows for easy sensitivity analysis, quickly identifying the most critical factors affecting system performance and guiding design optimization.
- **Practical Design Tool**: The formulas can be readily incorporated into design guidelines, handbooks, and software tools, providing engineers with a practical and efficient means of assessing lightning protection requirements.

However, the key challenge is to develop analytical formulas that are both simple and accurate. They must capture the essential physics of the problem without being overly complex or requiring extensive calibration. The formulas presented in the following subsections achieve this balance by leveraging the extensive simulation data generated using the rigorous 3D FEM and Monte Carlo approach (described in Sect. 6.4). The formulas are, in essence, distilled representations of the complex simulation results, capturing the key relationships in a compact and user-friendly form.

6.5.2 The Generalized Extreme Value (GEV) Distribution: A Statistical Foundation

The Generalized Extreme Value (GEV) distribution is a family of continuous probability distributions specifically designed to model the *extreme values* of a random variable. It is a generalization of the Gumbel, Fréchet, and Weibull distributions, and it is widely used in various fields dealing with extreme events, such as

- Hydrology (modeling maximum flood levels)
- Meteorology (modeling maximum wind speeds)
- Seismology (modeling maximum earthquake magnitudes)
- Finance (modeling extreme market losses)
- And, relevant to this chapter, lightning studies (modeling maximum induced voltages).

In the context of lightning-induced overvoltage, the GEV distribution is particularly appropriate because we are primarily concerned with the maximum induced voltage on the line, as this determines whether a flashover will occur. Previous researches [38–40] have demonstrated that the GEV distribution provides a good fit to the statistical distribution of peak induced voltages obtained from simulations and measurements.

The GEV distribution is characterized by three parameters:

- **Location Parameter (μ)**: This parameter represents the "center" or "typical" value of the extreme values. In the context of induced voltages, it would be related to the average magnitude of the peak induced voltages, influenced by factors like average lightning current and distance to the strike.
- **Scale Parameter (σ)**: This parameter represents the spread or variability of the extreme values. A larger scale parameter indicates a wider range of possible peak induced voltages, reflecting greater uncertainty or variability in the lightning parameters or environmental conditions.
- **Shape Parameter (ζ)**: This parameter determines the "tail behavior" of the distribution, which is crucial for characterizing the probability of very rare, but very high, voltage events. The shape parameter dictates whether the distribution is
  - **$\zeta < 0$ (Weibull type)**: The distribution has an upper bound. There is a maximum possible induced voltage, beyond which the probability is zero.
  - **$\zeta = 0$ (Gumbel type)**: The distribution is unbounded, but the probability of extremely high voltages decays exponentially.
  - **$\zeta > 0$ (Fréchet type)**: The distribution is unbounded, and the probability of extremely high voltages decays more slowly (following a power law).

The analytical formulas presented below are based on the assumption that the annual number of flashovers can be related to the cumulative distribution function (CDF) of the GEV distribution fitted to the induced voltage data. This is a reasonable assumption because the flashover probability is directly related to the probability of exceeding the insulator's withstand capability (as defined by the volt-time curve), which is, in turn, related to the distribution of extreme voltage values.

6.5.3 Formula for Mixed Ocean-Land Environments

For overhead distribution lines located near a coastline (the mixed ocean-land propagation path scenario, as discussed in Sect. 6.3.2.3), the following analytical formula is proposed to estimate the annual flashover rate (F_p) per 100 km of line (adapted from [18]):

$$F_P = 200N_g \times \left\{1 - \exp\left[-\left(1 + \varepsilon_{DS}\left(\frac{CFO - \mu_{DS}}{\sigma_{DS}}\right)\right)^{-1/\varepsilon_{DS}}\right]\right\} \tag{6.9}$$

where

- $\boldsymbol{F_p}$: The annual flashover rate per 100 km of line. This represents the expected number of flashovers per year per 100 km of line length due to *indirect* lightning strikes. The units are flashovers/(100 km-year).
- $\boldsymbol{N_g}$: The ground flash density (flashes/km^2/year). This is the number of lightning flashes striking the ground per square kilometer per year in the region of interest. This value is typically obtained from lightning location systems or from isokeraunic maps (as discussed in Chap. 1).
- **CFO**: The Critical Flashover Voltage of the insulator (in kV). This is the voltage level at which the insulator has a 50% probability of flashover under standard laboratory impulse conditions (as discussed in Sect. 6.2).
- $\boldsymbol{\varepsilon_{DS}, \mu_{DS}, \sigma_{DS}}$: These are the shape, location, and scale parameters of the GEV distribution, respectively. Crucially, these parameters are not constant; they are functions of the distance from the coastline (DS). This dependence on DS captures the complex influence of the ocean-land interface on the induced voltage distribution. The subscript "DS" emphasizes this dependence.

The dependence of ε_{DS}, μ_{DS}, and σ_{DS} on DS is modeled using the following empirical relationship (also from [18]):

$$X(DS) = \frac{A_{DS}}{\left(1 + e^{B_{DS}*(DS - C_{DS})}\right)} + D_{DS} \tag{6.10}$$

where

- **X(DS)** represents either ε_{DS}, μ_{DS}, or σ_{DS}. You apply this equation three times, once for each GEV parameter.
- **DS**: The distance from the coastline (in meters). This is the key independent variable that captures the influence of the ocean-land interface.
- A_{DS}, B_{DS}, C_{DS}, D_{DS}*:* These are constants that are specific to each of the GEV parameters (ε_{DS}, μ_{DS}, and σ_{DS}). These constants were determined empirically by fitting the analytical formula (Eq. 6.10) to the extensive simulation data generated using the 3D FEM and Monte Carlo approach (as described in Sect. 6.4 and in [18]. The fitting process involved finding the values of these constants that

Table 6.3 Constants for the mixed ocean -land scenario

	A_{DS}	B_{DS}	C_{DS}	D_{DS}
ε_{DS}	3.75081	1.35661	21.089	9.58087
μ_{DS}	11.7547	0.090677	18.9679	11.0785
σ_{DS}	16.552	0.122918	25.6815	45.0446

minimized the difference between the flashover rates predicted by the formula and the flashover rates obtained from the simulations, for a wide range of DS and CFO values.

Table 6.3 provides the constant values for calculating ε_{DS}, μ_{DS}, and σ_{DS} in the analytical formula for mixed ocean-land environments. These constants were determined through empirical fitting to 3D FEM simulation results.

Using the Formula (Step-by-Step)

1. **Obtain Input Parameters**
   (a) N_g: Determine the ground flash density for the region of interest (from lightning location systems or isokeraunic maps).
   (b) **CFO**: Obtain the CFO of the insulators used on the distribution line (from manufacturer data or standards).
   (c) **DS**: Measure or estimate the distance between the overhead line and the coastline.
2. **Calculate GEV Parameters**
   (a) Use Eq. 6.10 and the constants from Table 6.3 to calculate ε_{DS}, μ_{DS}, and σ_{DS}. You'll perform this calculation three times, once for each GEV parameter.
3. **Calculate F_p**
   (a) Substitute the calculated values of ε_{DS}, μ_{DS}, and σ_{DS} along with N_g and CFO, into the main formula (Eq. 6.9).
   (b) Make sure the term inside is greater than 0.
   (c) Evaluate the expression to obtain F_p, the annual flashover rate.

Interpret Result: The calculated F_p represents the expected number of flashovers per year per 100 km of line length, for the given line configuration, ground flash density, insulator CFO, and distance from the coastline.

6.5.3.1 Validation and Accuracy

The accuracy of this analytical formula has been rigorously validated by comparing its predictions to the results of detailed 3D FEM simulations coupled with Monte Carlo analysis, as described in Sect. 6.4 and in [18].

Figure 6.6 shows a comparison between the annual flashover rates predicted by the analytical formula and the results obtained from the full 3D FEM simulations. As can be seen, the agreement is excellent, demonstrating the accuracy of the proposed formula across a range of CFO values and distances from the coastline. The differences between the analytical formula and the simulation results are typically less than 1% for the parameter ranges considered in the study.

6.5.3.2 Limitations

It's important to acknowledge the limitations of this formula:

- **Specific Ground Conditions**: The formula is specifically developed for the mixed ocean-land scenario. The constants in Table 6.3 are derived from simulations that considered this specific ground configuration. The formula may not be directly applicable to other ground conditions (e.g., mountainous terrain, urban environments, or different soil stratification patterns).
- **Parameter Range**: The constants in Table 6.3 were derived from simulations with specific ranges of parameters (lightning current parameters, line height, soil resistivity on the land side). The formula's accuracy may be reduced outside of these ranges. Extrapolation beyond the validated range should be done with caution and, ideally, validated with additional simulations.
- **Simplified Line Model**: The formula is based on a simplified representation of the overhead line (single conductor, no detailed modeling of poles or grounding). While this simplification is generally valid for indirect strike analysis, it may not be appropriate for analyzing direct strikes or for situations where the detailed geometry of the line is crucial.
- **Vertical Channel**: This formula is based on the vertical channel.

Despite these limitations, the analytical formula provides a valuable tool for engineers, allowing for quick and accurate estimation of flashover rates in coastal areas without the need for computationally expensive simulations. It provides a practical means of assessing lightning protection requirements and optimizing design decisions, contributing to improved system reliability.

6.5.4 *Formula for Horizontally Stratified Soil*

For overhead distribution lines located above horizontally stratified soil (the two-layer model described in Sect. 6.3.2.2), a similar analytical formula is proposed (adapted from [16]):

$$F_{\mathrm{P}} = 200N_{\mathrm{g}} \times \left\{1 - \exp\left[-\left(1 + \varepsilon_H\left(\frac{\mathrm{CFO} - \mu_H}{\sigma_H}\right)\right)^{-1/\varepsilon_H}\right]\right\} \tag{6.11}$$

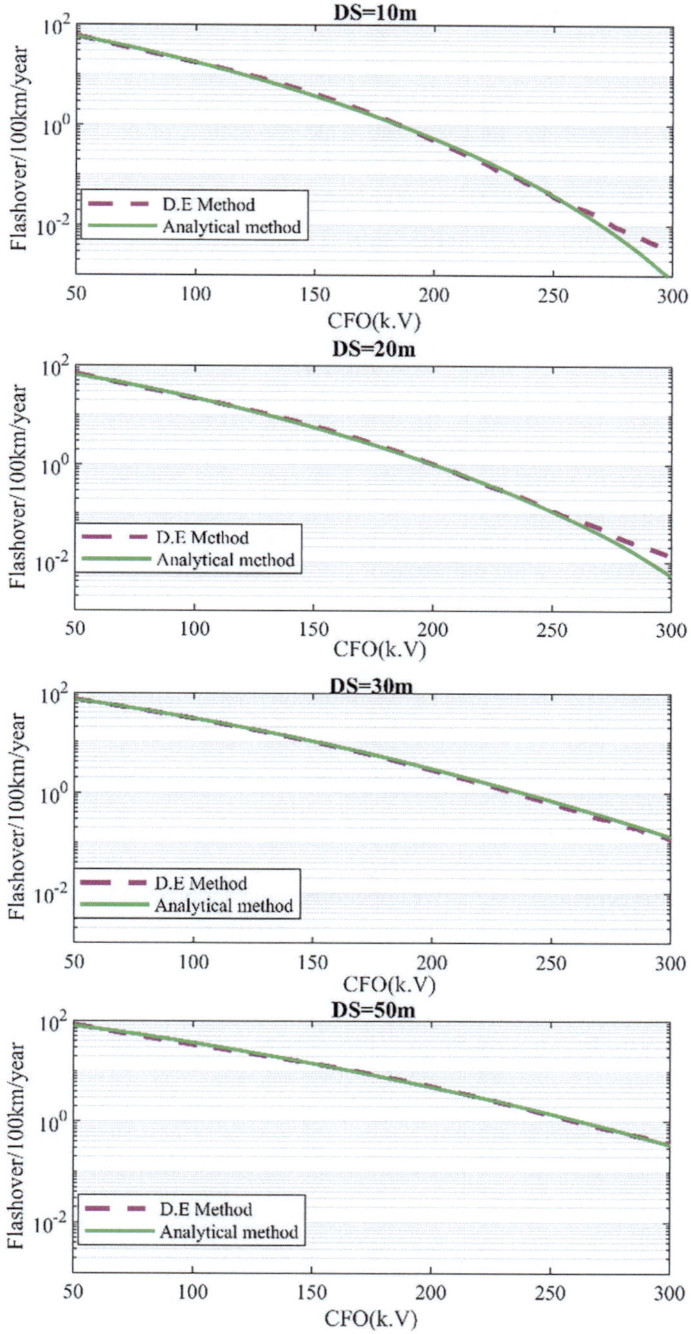

Fig. 6.6 Comparison of annual flashover rates predicted by the analytical formula (Eq. 6.9) and the 3D FEM/Monte Carlo simulations for a mixed ocean-land environment. Results are shown for various distances from the coastline (DS)

where

- F_p, N_g, and CFO have the same definitions as in the previous formula (annual flashover rate, ground flash density, and critical flashover voltage, respectively).
- ε_H, μ_H, σ_H: The shape, location, and scale parameters of the GEV distribution, respectively. These parameters are now functions of the depth of the upper soil layer (H). The subscript "H" emphasizes this dependence.

The dependence of ε_H, μ_H, and σ_H on H is modeled using the following empirical relationship (also from [16]):

$$X(H) = \frac{A_H}{B_H + H} + C_H \tag{6.12}$$

where

- $X(H)$ represents either ε_H, μ_H, or σ_H. Can apply this equation three times, once for each GEV parameter.
- H: The depth of the upper soil layer (in meters). This is the key independent variable that captures the influence of soil stratification.
- A_H, B_H, C_H: These are constants that are specific to each of the GEV parameters (ε_H, μ_H, σ_H) and to the specific combination of upper- and lower-layer conductivities. These constants were determined empirically by fitting the analytical formula (Eq. 6.12) to the extensive simulation data generated using the 3D FEM and Monte Carlo approach (as described in Sect. 6.4 and in [16]).

Table 6.4 (adapted from [5, 16]) provides the values of these constants for two different scenarios:

1. **Scenario 1**: Upper layer conductivity (σ_u) = 0.001 S/m, lower layer conductivity (σ_d) = 0.01 S/m (less conductive upper layer over a more conductive lower layer).
2. **Scenario 2**: Upper layer conductivity (σ_u) = 0.01 S/m, lower layer conductivity (σ_d) = 0.001 S/m (more conductive upper layer over a less conductive lower layer).

Using the Formula (Step-by-Step)

Table 6.4 Constant values for calculating ε_H, μ_H, and σ_H in the analytical formula for horizontally stratified soil: (a) $\sigma_u = 0.001$ S/m, $\sigma_d = 0.01$ S/m, (b) $\sigma_u = 0.01$ S/m, $\sigma_d = 0.001$ S/m

Scenario	Parameter	A_H	B_H	C_H
$\sigma_u = 0.001$ S/m	ε_H	0.82	2.5	– 0.165
	μ_H	– 176.2	18.55	15
	σ_H	– 1050	18.32	85
$\sigma_u = 0.01$ S/m	ε_H	13.15	200	– 0.119
	μ_H	479	133	– 0.81
	σ_H	4.9	– 53.12	38

1. **Obtain Input Parameters**
   (a) N_g: Determine the ground flash density for the region.
   (b) CFO: Obtain the CFO of the insulators.
   (c) H: Determine (or estimate) the depth of the upper soil layer.
   (d) σ_u and σ_d: Determine the conductivities of the upper and lower soil layers.
2. **Select Constants**: Choose the appropriate set of constants (A_H, B_H, C_H) from Table 6.4 based on the conductivities of the upper and lower layers (Scenario 1 or Scenario 2).
3. **Calculate GEV Parameters**: Use Eq. 6.12 and the selected constants to calculate ε_H, μ_H, and σ_H. You'll perform this calculation three times, once for each GEV parameter.
4. **Calculate F_p**: Substitute the calculated values of ε_H, μ_H, and σ_H along with N_g and CFO, into the main formula (Eq. 6.11).
5. **Interpret Result**: The calculated F_p represents the expected number of flashovers per year per 100 km of line length, for the given line configuration, ground flash density, insulator CFO, and soil stratification parameters (H, σ_u, σ_d).

Validation and Accuracy

The accuracy of this analytical formula, like the one for the ocean-land interface, has been rigorously validated by comparing its predictions to the results of detailed 3D FEM simulations coupled with Monte Carlo analysis, as described in Sect. 6.4 and documented in [5, 18].

Figure 6.7 shows a comparison between the annual flashover rates predicted by the analytical formula and the results obtained from the full 3D FEM simulations for various upper layer depths and conductivity combinations. The agreement between the analytical formula and the simulation results is excellent, confirming the formula's validity and accuracy within the parameter ranges considered in the study. The differences are typically small, demonstrating the formula's ability to capture the complex effects of soil stratification on flashover rates.

Limitations

It's important to acknowledge the limitations of this formula:

- **Specific Ground Model**: The formula is specifically developed for the two-layer horizontal soil model. It may not be directly applicable to other ground configurations, such as
  - Multi-layered soil with more than two layers.
  - Vertically stratified soil.
  - Soil with continuously varying conductivity.
  - Significant terrain features (hills, valleys) are present.
- **Parameter Range**: The constants in Table 6.4 were derived from simulations with specific ranges of parameters (lightning current parameters, line height, soil conductivities). The formula's accuracy may be reduced outside of these ranges.

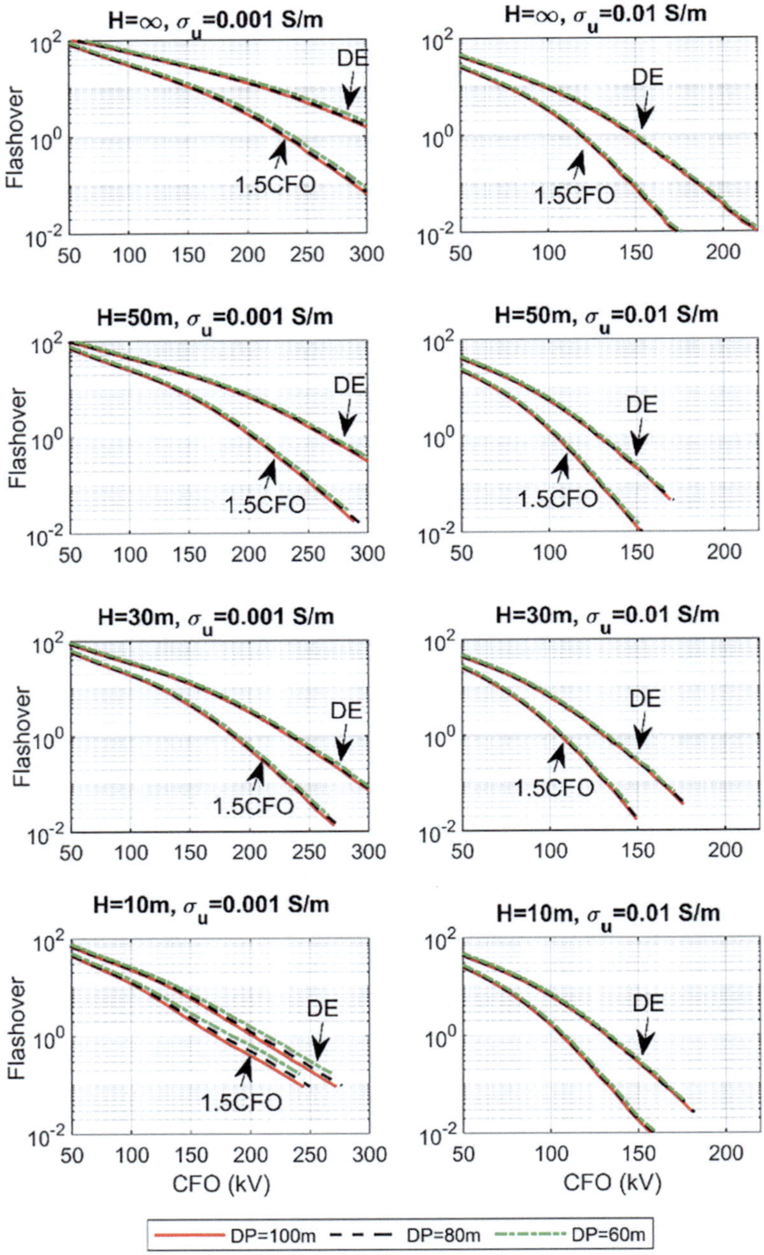

Fig. 6.7 Comparison of annual flashover rates predicted by the analytical formula (Eq. 6.11) and the 3D FEM/Monte Carlo simulations for horizontally stratified soil. Results are shown for various upper layer depths (H) and conductivity combinations (σ_u and σ_d)

Extrapolation beyond the validated range should be done with caution and, ideally, validated with additional simulations or field data.

- **Simplified Line Model**: The formula is based on a simplified representation of the overhead line (single conductor, no detailed modeling of poles or grounding). While this simplification is generally valid for indirect strike analysis, it may not be appropriate for analyzing direct strikes or for situations where the detailed geometry of the line is crucial.
- **Vertical Channel**: This formula is based on the vertical channel.

Conclusion of Sect. 6.5:

The two analytical formulas presented in this section—one for mixed ocean-land environments and one for horizontally stratified soil—provide engineers with powerful, practical tools for estimating lightning-induced flashover rates on overhead distribution lines. These formulas, derived from extensive 3D FEM simulations and validated against those simulations, offer a significant improvement in accuracy and efficiency compared to traditional methods like the 1.5CFO criterion. They allow for rapid assessment of flashover risk, facilitating informed decisions regarding lightning protection design and mitigation strategies. They bridge the gap between complex numerical simulations and practical engineering applications, enabling engineers to quickly evaluate the impact of various parameters (distance from coastline, soil layer depth, and conductivity) without the need for time-consuming simulations.

Now, let's move on to Sect. 6.6 (Mitigation Strategies). Since we've already drafted a substantial portion of this section, the complete Sect. 6.6 is provided here, incorporating your previous feedback and ensuring consistency.

6.6 Mitigation Strategies

The preceding sections have established the significant threat posed by indirect lightning strikes to overhead distribution lines and have presented advanced methods for accurately predicting flashover rates. However, understanding the risk is only the first step. Effective mitigation strategies are crucial for protecting distribution systems from lightning-induced damage and ensuring reliable power delivery. This section will discuss the primary mitigation techniques, emphasizing how the insights gained from the advanced modeling approaches (Sects. 6.4 and 6.5) inform the optimal design and implementation of these strategies. We will primarily focus on techniques relevant to indirect lightning strikes, although some methods (e.g., shield wires) also offer protection against direct strikes.

As discussed in Chap. 2, the primary mitigation strategies include surge arresters, shield wires, and grounding systems. This section also includes a discussion on the selection of insulators.

6.6.1 Surge Arresters: The First Line of Defense

Surge arresters, also known as surge protective devices (SPDs), are the primary defense against lightning-induced overvoltage on distribution lines. Their function is to limit the voltage on the line to a safe level, preventing flashover of insulators and damage to connected equipment. As Chap. 2 described in detail, surge arresters operate by providing a low-impedance path to ground for surge currents, diverting the energy away from the line and connected equipment.

- **Operating Principle**: Under normal operating conditions, a surge arrester presents a high impedance, effectively acting as an open circuit. When an overvoltage exceeding the arrester's clamping voltage appears on the line, the arrester rapidly transitions to a low impedance state, providing a path for the surge current to flow to ground. Once the overvoltage is discharged and the voltage returns to normal levels, the arrester returns to its high-impedance state.
- **Key Characteristics:**
  - **Clamping Voltage**: The voltage level at which the arrester begins to conduct significant current. This should be chosen to be above the normal operating voltage of the line (with a suitable margin), but below the CFO of the insulators.
  - **Energy Rating**: The amount of energy (in Joules) the arrester can safely absorb without being damaged. This should be selected based on the expected lightning current magnitudes and durations in the region, and informed by the modeling techniques presented in this chapter.
  - **Response Time**: The time it takes for the arrester to transition from its high-impedance state to its low-impedance state. Faster response times are desirable for protecting against fast-rising lightning surges.
  - **Discharge Current Capability**: The maximum current the arrester can safely conduct to ground.
- **Types of Surge Arresters**: Chap. 2 provides a comprehensive overview of various surge arrester technologies. While several types exist, metal-oxide varistors (MOVs), particularly those based on zinc oxide (ZnO), are the most widely used in modern distribution systems due to their excellent performance characteristics:
  - **Nonlinear V-I Characteristic**: ZnO MOVs exhibit a highly nonlinear voltage-current (V-I) characteristic. This means their resistance decreases dramatically as the voltage increases, providing a very low impedance path for surge currents, effectively clamping the voltage.
  - **Fast Response Time**: MOVs respond very quickly to overvoltages (typically within nanoseconds), effectively clamping fast-rising lightning surges.
  - **High Energy Absorption Capability**: MOVs can absorb significant amounts of energy without being damaged, making them suitable for handling the high energy associated with lightning strikes.

  - **Long Service Life**: MOVs have a long service life under normal operating conditions, although repeated exposure to high-energy surges can lead to degradation.

- **Placement and Spacing**: The placement and spacing of surge arresters along the distribution line are critical for adequate protection. As demonstrated by the simulations and analytical formulas presented in this chapter and in [16, 18, 26], the induced voltage varies along the line, depending on factors such as ground conditions and distance from the striking point. Therefore, surge arresters must be strategically placed to provide adequate protection along the entire length of the line.

  - **Ideal Scenario**: Ideally, surge arresters would be installed on every pole to provide the maximum level of protection. However, this is often economically impractical.
  - **Practical Considerations**: In practice, a balance must be struck between protection level and cost. The optimal spacing depends on several factors, including:

    Line CFO: Lines with lower CFO insulators require more frequent arrester placement.
    Ground Flash Density: Areas with higher lightning activity require more frequent arrester placement.
    Soil Resistivity and Ground Conditions: As demonstrated throughout this chapter, areas with high soil resistivity, stratified soil, or proximity to coastlines require more careful consideration of arrester placement, as these conditions can lead to higher induced voltages. The analytical formulas presented in Sect. 6.5 can be used to inform these decisions, providing a quantitative basis for optimizing arrester spacing.
    Presence of Shield Wires: If shield wires are present, they can reduce the induced voltages, potentially allowing for greater spacing between arresters.

  - **IEEE Std 1410**: The IEEE Std 1410 [7] provides guidelines for improving the lightning performance of distribution lines, including recommendations for surge arrester placement. However, it's crucial to recognize that these guidelines are based on simplified models and may not be adequate for all scenarios, particularly those involving complex ground conditions. The advanced modeling techniques and analytical formulas presented in this chapter provide a more accurate basis for optimizing arrester placement in these challenging environments.
  - **Coordination with Other Protective Devices**: Surge arresters should be properly coordinated with other protective devices in the system, such as fuses and circuit breakers. This coordination ensures that the arresters operate correctly to limit overvoltage's, while also preventing unnecessary tripping of protective devices due to the arrester's conduction during a surge

6.6.2 *Shield Wires: Intercepting Direct Strikes and Reducing Induced Voltages*

Shield wires (also known as overhead ground wires) are conductors installed above the phase conductors of a distribution line, typically connected to the grounded support structures (poles or towers). While their primary function is to intercept direct lightning strikes and prevent them from reaching the phase conductors (as discussed in Chap. 2), shield wires also provide a degree of protection against indirect strikes by reducing the induced voltages on the phase conductors.

- **Mechanism of Shielding**: Shield wires reduce induced voltages through two primary mechanisms:
  - **Electrostatic Shielding**: The presence of the grounded shield wire alters the electric field distribution around the phase conductors. This reduces the capacitive coupling between the lightning channel and the line, leading to lower induced voltages.
  - **Electromagnetic Shielding**: The shield wire provides a low-impedance path to ground for currents induced by the changing magnetic field associated with the lightning strike. This reduces the inductive coupling between the lightning channel and the phase conductors.

  The shield wire effectively "intercepts" a portion of the electromagnetic field, diverting the energy to ground.
- **Effectiveness of Shielding:** The effectiveness of shield wires in reducing induced voltages depends on several factors:
  - **Shield Wire Height and Position**: The height of the shield wire above the phase conductors and its horizontal position relative to the phase conductors are critical. Optimal placement is determined through detailed electromagnetic analysis, often using the electro geometric model (as discussed in Chap. 1) or more advanced numerical techniques. The goal is to maximize the interception of direct strikes while also minimizing the coupling between the lightning channel and the phase conductors for indirect strikes.
  - **Number of Shield Wires:** Multiple shield wires provide better shielding than a single wire, further reducing the induced voltages.
  - **Grounding Resistance**: Low grounding resistance at the poles is *essential* for effective shield wire performance. The shield wire must provide a low-impedance path to ground for both the intercepted direct strike currents and the induced currents. High grounding resistance can negate the benefits of the shield wire.
  - **Lightning Current Characteristics**: The effectiveness of shielding is also influenced by the lightning current parameters (peak current, rise time). Faster-rising currents tend to induce higher voltages, making shielding more challenging.

- **Combination with Surge Arresters**: While shield wires can significantly reduce induced voltages, they cannot completely eliminate them, particularly for nearby strikes. Therefore, shield wires are typically used in conjunction with surge arresters to provide a comprehensive lightning protection system. The shield wires reduce the frequency and magnitude of the overvoltage's that reach the surge arresters, extending their lifespan and improving overall system reliability. The combination of shield wires and surge arresters provides a layered defense, addressing both direct and indirect lightning threats.

6.6.3 Grounding Systems: Providing a Low-Impedance Path to Earth

Effective grounding is essential for both direct and indirect lightning protection.

A low-impedance grounding system provides a path for lightning currents to dissipate safely into the earth, minimizing voltage rises and preventing damage to equipment and ensuring personnel safety. The importance of grounding is particularly critical in the context of the complex ground conditions discussed in this chapter (stratified soil, ocean-land interfaces).

- **Grounding System Objectives**
  - **Minimize Ground Potential Rise (GPR)**: As discussed in Sect. 6.3.2.1, lightning strikes to ground cause a significant rise in ground potential. A low-impedance grounding system helps to limit this GPR, reducing the stress on insulators and equipment connected to the grounding system.
  - **Provide a Low-Impedance Path for Surge Currents**: The grounding system must provide a low-impedance path for both directly intercepted currents (from shield wires) and induced currents (from surge arresters) to flow to ground. This allows the surge energy to dissipate safely into the earth, preventing it from damaging the power system.
  - **Ensure Personnel Safety**: A well-designed grounding system helps to minimize step and touch potentials, protecting personnel from dangerous voltage gradients during lightning events.
- **Grounding Electrode Types**: Various types of grounding electrodes are used in distribution systems, including:
  - **Ground Rods**: Vertical rods driven into the earth. These are the most common type of grounding electrode, particularly for distribution poles. The length and number of rods are chosen based on soil resistivity and the desired grounding resistance.
  - **Ground Grids**: A network of interconnected conductors buried in the earth, typically used at substations. Grids provide a large surface area for current dissipation and are effective in reducing GPR.

  - **Counterpoise Conductors**: Horizontal conductors buried in the earth, often used in conjunction with shield wires to provide a continuous low-impedance path along the line. Counterpoise conductors can be particularly effective in high-resistivity soils.
  - **Chemical Electrodes**: Electrodes filled with conductive materials (e.g., bentonite clay, specialized grounding compounds) to improve ground contact in high-resistivity soils. These materials help to lower the effective resistivity of the soil immediately surrounding the electrode.

- **Grounding System Design Considerations:**

  - **Soil Resistivity**: The design of the grounding system must take into account the soil resistivity at the site. As extensively discussed in this chapter, soil resistivity is a critical factor influencing the performance of the grounding system. In areas with high soil resistivity, more extensive grounding systems (e.g., longer rods, larger grids, chemical electrodes) may be required to achieve a sufficiently low ground impedance. The variation of soil resistivity with depth (stratification) and location (ocean-land interface) must also be considered.
  - **Ground Stratification**: As demonstrated in Sect. 6.3.2.2 and in the research presented in [16], ground stratification significantly affects the propagation of lightning currents and the resulting GPR. The grounding system design should consider the presence of layers with different conductivities. For example, if a highly resistive layer overlies a more conductive layer, deeper grounding electrodes that penetrate into the lower layer may be more effective. The analytical formula presented in Sect. 6.5.4 provides a valuable tool for optimizing grounding system design in stratified soil.
  - **Ocean-Land Interface**: In coastal areas, the high conductivity of seawater significantly alters the ground potential distribution, as discussed in Sect. 6.3.2.3. Grounding systems in these areas require careful design to account for the mixed propagation path. The analytical formula presented in Sect. 6.5.3 provides a valuable tool for optimizing grounding system design in these complex environments.
  - **Dynamic Grounding**: In some cases, dynamic grounding techniques may be employed. These techniques involve actively adjusting the grounding impedance based on real-time measurements of soil conditions (e.g., soil moisture content, temperature). This can be particularly useful in areas with significant seasonal variations in soil resistivity.
  - **Coordination with Surge Arresters**: The design of grounding and surge arrester should be coordinate.

- **Standards and Guidelines**: Several standards and guidelines provide recommendations for grounding system design, including IEEE Std 80 ("IEEE Guide for Safety in AC Substation Grounding") [41] and IEC 62305 ("Protection against lightning") [42]. However, it's important to recognize that these standards may not fully address the complexities of non-uniform ground conditions.

The advanced modeling techniques and analytical formulas presented in this chapter provide a more accurate basis for optimizing grounding system design in these challenging scenarios.

6.6.4 Insulator Considerations

Overhead distribution lines rely on insulators to electrically isolate the energized conductors from the grounded support structures. The choice of insulator type, its electrical characteristics, and its coordination with other protective devices are crucial for preventing flashover.

- **Insulator Types**
  - **Porcelain Insulators**: Traditional, mechanically strong, and relatively inexpensive, but susceptible to surface contamination.
  - **Glass Insulators**: Similar to porcelain, with slightly better contamination resistance in some cases.
  - **Polymer (Composite) Insulators**: Increasingly popular due to their hydrophobicity (water repellency), lighter weight, and, in some cases, higher mechanical strength. However, long-term aging and degradation mechanisms are still
- **Critical Flashover Voltage (CFO)**: As defined in Sect. 6.2, the Critical Flashover Voltage (CFO) is the voltage level at which an insulator has a 50% probability of flashover under standard laboratory impulse conditions (typically a 1.2/50 μs waveshape). The CFO is a key parameter provided by insulator manufacturers, based on standardized testing (e.g., IEC 60060-1 [2]). It serves as a baseline for insulation coordination.
- **Volt-Time Characteristic:** As emphasized throughout this chapter, the CFO alone is insufficient for accurate flashover prediction under realistic lightning conditions. The volt-time characteristic of the insulator, which describes the relationship between the applied voltage and the time to flashover, provides a much more complete picture of the insulator's behavior. The Volt-Time Disruptive Model (Sect. 6.2.2) utilizes this characteristic to account for the dynamic response of the insulator and the shape of the induced voltage waveform. The parameters for the volt-time model (U_0, K, and DE^*) are determined through laboratory testing, typically involving a range of impulse voltage waveforms.
- **Insulation Coordination**: The process of insulation coordination involves selecting insulators with appropriate CFO values (and volt-time characteristics) to ensure that the probability of flashover is acceptably low, given the expected overvoltage's on the line. This selection must consider:
  - **Expected Over voltages**: The magnitude and waveshape of the expected lightning-induced voltages, calculated using the advanced modeling techniques described in Sect. 6.4, and taking into account the specific ground conditions (Sect. 6.3.2) and lightning current parameters (Sect. 6.3.1).

- **Presence of Protective Devices**: The presence of surge arresters and shield wires, which reduce the magnitude and frequency of overvoltage's reaching the insulators. The coordination between insulators and protective devices is crucial.
- **Environmental Conditions**: Factors such as pollution, humidity, and altitude can affect the insulator's withstand capability. For example, high altitude reduces the air density, decreasing the breakdown voltage. Pollution and humidity can create conductive paths on the insulator surface, also reducing the flashover voltage.
- **Economic Considerations**: Higher CFO insulators typically have higher costs. The optimal choice involves balancing performance (low flashover probability) and cost. A risk-based approach is often used, considering the cost of outages and the consequences of failures.

- **Impact of Ground Conditions**: The findings presented in this chapter, particularly regarding the influence of ground stratification (Sect. 6.3.2.2) and the ocean-land interface (Sect. 6.3.2.3), have significant implications for insulation coordination. Traditional methods that rely solely on the 1.5CFO criterion and assume homogeneous ground may significantly underestimate the flashover risk in these complex environments. The use of the Volt-Time Disruptive Model, combined with accurate induced voltage calculations (considering the specific ground conditions), allows for a more precise assessment of the required insulator strength and a more reliable design. This avoids both under-protection (leading to excessive flashovers) and over-protection (using unnecessarily expensive insulators).

6.6.5 Summary of Mitigation Strategies

Effective lightning protection of overhead distribution lines requires a holistic, multi-layered approach that incorporates the following key elements, all working in a coordinated manner:

- **Surge Arresters**: Strategically placed along the line to limit overvoltage's due to both direct and indirect strikes.
  - The selection (type, clamping voltage, energy rating) and placement (spacing) should be informed by accurate induced voltage calculations, considering the specific ground conditions, lightning environment, and the statistical distribution of lightning current parameters (as discussed in Chaps. 4 and 5). The methods presented in this chapter, including the analytical formulas in Sect. 6.5, are crucial for optimizing surge arrester placement.
- **Shield Wires:** Used to intercept direct strikes and reduce induced voltages from indirect strikes. The design (height, position, number of wires) should be optimized based on electromagnetic analysis (e.g., using the electro geometric model,

as discussed in Chap. 1, or more advanced numerical techniques) and consider the grounding system.

- **Grounding Systems**: Designed to provide a low-impedance path to earth for lightning currents, minimizing GPR and ensuring the effective operation of surge arresters and shield wires. The design must account for soil resistivity, stratification (using models like those in Sect. 6.3.2.2), and the presence of any ocean-land interface (Sect. 6.3.2.3). The analytical formulas presented in Sect. 6.5 provide a valuable tool for optimizing grounding system design in these complex environments.
- **Insulation Coordination**: Selecting insulators with appropriate CFO values and utilizing the Volt-Time Disruptive Model (Sect. 6.2.2) for more accurate flashover prediction. This allows for a more robust and cost-effective design, avoiding both under-protection (leading to excessive flashovers) and over-protection (using unnecessarily expensive insulators). The volt-time approach, coupled with accurate induced voltage calculations, is essential for proper insulation coordination, especially in challenging environments.

The advanced modeling techniques (3D FEM, Monte Carlo simulation) and analytical formulas presented in this chapter provide engineers with the tools to move beyond simplified, conservative design approaches and to implement optimized lightning protection strategies that are tailored to the specific conditions of each distribution line. This leads to improved system reliability, reduced outage rates, enhanced safety, and optimized cost-effectiveness. The insights from this chapter directly inform the selection, placement, and coordination of these mitigation measures.

6.7 Case Studies

The theoretical concepts, modeling techniques, and analytical formulas presented in the preceding sections have significant practical implications for the design and protection of overhead [5, 16, 18, 26, 43], to illustrate these applications and to highlight the importance of considering complex ground conditions and the volt-time characteristics of insulators.

6.7.1 Impact of the Ocean-Land Interface

The research presented in [24, 26, 43] extensively investigated the influence of the ocean-land interface on lightning-induced voltages and flashover rates. This scenario is particularly relevant for distribution lines located in coastal regions and for offshore wind farms, where the high conductivity of seawater significantly alters the electromagnetic field propagation compared to a homogeneous land environment.

- **Scenario**: A typical overhead distribution line configuration was considered, with a conductor height of 10 m above ground. The key variable was the distance (DS) between the line and the coastline, ranging from close proximity (e.g., 10 m) to distances where the influence of the ocean becomes negligible (effectively representing a homogeneous land environment). The soil on the land side was assigned a conductivity of 0.001 S/m. Lightning strikes were modeled using the techniques described in Sect. 6.4, with statistically representative lightning current parameters (as detailed in Chap. 4). Both the 1.5CFO criterion and the Volt-Time Disruptive Model were used to assess flashover.
- **Key Findings**
  - **Significant Underestimation by 1.5CFO**: The simulations consistently demonstrated that the 1.5CFO criterion significantly underestimated the annual flashover rate compared to the more accurate Volt-Time Disruptive Model. This underestimation was particularly pronounced at larger distances from the coastline. For instance, at a CFO of 200 kV, the 1.5CFO criterion underestimated the flashover rate by 76–94% for various DS values.
  - **Complex Relationship with DS**: The annual flashover rate did not vary monotonically with DS. The interaction between the reduced attenuation over seawater, reflections at the coastline, and the altered GPR created a complex relationship.
  - **Waveshape Distortion**: The presence of the ocean-land interface significantly altered the waveshape of the induced voltages, with faster rise times and, in some cases, multiple peaks. This highlights the importance of using the volt-time model, which accounts for waveshape effects.
  - **Validation of Analytical Formula**: The analytical formula presented in Sect. 6.5.3, specifically developed for the ocean-land scenario, was shown to provide excellent agreement with the detailed 3D FEM simulation results, validating its accuracy and usefulness for practical applications. Refer to Fig. 6.1 here, showing the comparison between the 1.5CFO, Volt-Time model, and analytical formula for the ocean-land scenario.

- **Practical Implications**: This case study clearly demonstrates that neglecting the influence of the ocean-land interface in lightning protection design for coastal distribution lines can lead to a significant underestimation of the flashover risk. Traditional design methods based on the 1.5CFO criterion and homogeneous ground assumptions are inadequate in these environments. The use of advanced modeling techniques (like 3D FEM) or the analytical formula presented in Sect. 6.5.3 is essential for accurate risk assessment and the proper selection and placement of protective devices (surge arresters, shield wires) and appropriate insulator selection.

6.7.2 *Influence of Horizontally Stratified Soil*

Reference [16] investigated the impact of horizontally stratified soil on lightning-induced flashovers. This scenario is common in many regions, where soil properties vary with depth due to differences in soil composition, moisture content, or geological formations.

- **Scenario**: An overhead distribution line was modeled above a two-layer horizontal soil model. The key parameters were the depth of the upper layer (H) and the conductivities of the upper (σ_u) and lower (σ_d) layers. Simulations were performed using 3D FEM, varying H and considering different combinations of σ_u and σ_d. Both the 1.5CFO criterion and the Volt-Time Disruptive Model were used for flashover assessment.
- **Key Findings**
  - **Significant Impact of Stratification**: The simulations showed that the presence of soil stratification significantly affected the induced voltage waveforms and the resulting flashover rates.

    The assumption of a homogeneous soil could lead to either underestimation or overestimation of the risk, depending on the specific layer conductivities and depth.
  - **Underestimation by 1.5CFO**: As with the ocean-land scenario, the 1.5CFO criterion consistently underestimated the flashover rate compared to the Volt-Time Disruptive Model. The degree of underestimation varied depending on the specific soil parameters, but it could be substantial (up to 85% in some cases).
  - **Complex Dependence on H, σ_u, and σ_d**: The flashover rate exhibited a complex, non-monotonic dependence on the upper layer depth (H) and the conductivity ratio (σ_u/σ_d). Different combinations of these parameters led to different electromagnetic field propagation patterns and, consequently, different induced voltage characteristics.
  - **Validation of Analytical Formula**: The analytical formula presented in Sect. 6.5.4, specifically developed for the two-layer horizontal soil scenario, was shown to provide excellent agreement with the detailed 3D FEM simulation results, validating its accuracy and usefulness for practical applications. Figure 6.2 shows the comparison between the 1.5CFO, Volt-Time model, and analytical formula for the stratified soil scenario.

    - **Practical Implications**: This case study highlights the importance of considering soil stratification in lightning protection design. Assuming a homogeneous soil with an "average" resistivity can lead to *inaccurate* risk assessments and potentially inadequate protection. The use of the two-layer soil model, combined with either 3D FEM simulations or the analytical formula (Sect. 6.5.4), allows for a more realistic representation of the ground and more accurate flashover predictions. This information is crucial for selecting appropriate insulator CFO values, optimizing surge arrester placement, and designing effective grounding systems.

6.7.3 The Effect of Striking Distance Calculation

References [16, 18, 26] for showing that the striking distance models has effect on the number of yearly flashovers.

6.7.4 The 2022 Texas Coastal Grid Collapse (Brief Mention)

The 2022 Texas Coastal Grid Collapse provides a real-world example of the devastating consequences of widespread power outages, some of which were attributed to lightning-induced events.

While this event involved a complex combination of factors (including extreme weather, generation failures, and grid management issues), the post-event analysis highlighted several contributing factors directly related to lightning protection:

- **Insufficient Surge Arrester Coverage**: A significant percentage of transmission towers lacked surge arresters, leaving them vulnerable to overvoltage. This underscores the critical importance of adequate surge arrester deployment, as discussed in Sect. 6.6.1.
- **Soil Ionization**: High soil moisture content from antecedent rainfall transiently reduced footing impedance, potentially increasing the risk of back flashover at some towers. This highlights the influence of ground conditions (specifically, soil moisture and its impact on conductivity) on lightning protection performance, as discussed in Sect. 6.3.2.
- **Relay Miscoordination**: Overcurrent relays did not coordinate properly with surge arrester operation, delaying fault isolation and contributing to cascading outages. This emphasizes the need for proper system coordination and the importance of accurate modeling to understand the interaction between different protective devices.

While not solely attributable to lightning, the Texas event serves as a stark reminder of the vulnerability of power systems to extreme weather events and the critical need for robust lightning protection measures, informed by accurate modeling and risk assessment. The advanced techniques presented in this chapter contribute to addressing these needs.

6.7.5 Summary of Case Studies

These case studies, drawn from the research presented in this book and real-world events, illustrate the practical application and value of the advanced modeling techniques and analytical formulas developed in this chapter. They demonstrate the following key points:

- **Ground conditions (ocean-land interface, soil stratification) have a significant and complex impact on lightning-induced flashover rates**. Ignoring these factors, as is often done in simplified analyses, can lead to substantial underestimation of the risk.
- **The Volt-Time Disruptive Model provides a more accurate assessment of flashover probability than the traditional 1.5CFO criterion**, particularly in complex environments where the induced voltage waveforms deviate significantly from the standard impulse.
- **The analytical formulas presented in** Sect. 6.5 **offer a practical and efficient way to estimate flashover rates in areas with complex ground conditions, without requiring extensive numerical simulations**. These formulas can be readily used by engineers for design and risk assessment.
- **Real-world events, like the Texas Coastal Grid Collapse, underscore the importance of comprehensive lightning protection and accurate risk assessment**. The advanced modeling techniques presented in this chapter contribute to improving the resilience of power systems against lightning-induced outages.

By incorporating these insights into the design, operation, and maintenance of distribution systems, engineers can significantly improve system reliability, reduce outage rates, enhance safety, and minimize the economic impact of lightning strikes.

6.8 Chapter Summary

This chapter has presented a comprehensive methodology for accurately predicting the probability of flashover on overhead distribution lines due to *indirect* lightning strikes, building upon the foundational knowledge established in Chaps. 1–5. The research presented here, and detailed in [16, 18, 26], makes significant advancements over traditional approaches by

- **Incorporating the Volt-Time Disruptive Model**: Replacing the simplistic 1.5CFO criterion with the physically realistic volt-time model, which accounts for the dynamic behavior of insulators and the influence of the *entire* induced voltage waveform, not just its peak value.
- **Accounting for Complex Ground Conditions**: Moving beyond the idealized assumption of a homogeneous, perfectly conducting ground, and rigorously modeling the significant impact of:

  - **Horizontally Stratified Soil**: Using a two-layer model to represent variations in soil conductivity with depth.
  - **Mixed Ocean-Land Propagation Paths**: Accounting for the drastically different electrical properties of seawater and land in coastal regions.

- **Employing Advanced Numerical Techniques**: Utilizing 3D Finite Element Method (FEM) simulations in COMSOL Multiphysics to accurately solve Maxwell's equations and calculate induced voltages in these complex environments.
- **Leveraging Statistical Methods**: Employing Monte Carlo simulations to incorporate the inherent randomness of lightning parameters (strike location, peak current, rise time, etc.) and to obtain statistically robust estimates of annual flashover rates.
- **Developing Practical Analytical Formulas**: Deriving and validating simplified analytical formulas for flashover rate estimation in both stratified soil and ocean-land scenarios, providing engineers with efficient tools for design and risk assessment.

The key findings of this research, supported by extensive simulations and comparisons with traditional methods, can be summarized as follows:

- **The 1.5CFO criterion consistently underestimates the flashover rate due to indirect lightning strikes, particularly in the presence of non-uniform ground conditions**. This underestimation can be substantial (ranging from 60% to over 90% in some of the scenarios investigated), potentially leading to under-designed lightning protection systems and increased outage risk.
- **The Volt-Time Disruptive Model provides a significantly more accurate prediction of flashover probability**. By considering the dynamic response of the insulator and the entire induced voltage waveform, it captures the essential physics of the breakdown process.
- **Ground conditions have a dominant influence on induced voltages and flashover rates**. Soil resistivity, stratification, and the presence of an ocean-land interface dramatically alter the propagation of electromagnetic fields and the resulting voltage waveforms. These effects cannot be ignored in an accurate lightning protection design.
- **The distance from the coastline (DS) in ocean-land environments, and the depth (*H*) and conductivity of the upper layer in stratified soil scenarios, are critical parameters that significantly influence flashover risk.**
- **Pole spacing (DP) indirectly affects the overall flashover probability of a line.** Closer spacing, while potentially reducing the magnitude of induced voltage at individual insulators, increases the number of potential flashover points, leading to a higher overall line flashover rate.
- **The analytical formulas presented in** Sect. 6.5 **provide accurate and efficient tools for estimating flashover rates in the specific scenarios of mixed**

ocean-land propagation and horizontally stratified soil. These formulas, validated against detailed 3D FEM simulations, offer a practical alternative to computationally expensive numerical methods for many engineering applications.

The case studies presented in Sect. 6.7 further illustrated the practical implications of these findings, demonstrating how the advanced modeling techniques and analytical formulas can be applied to real-world scenarios to improve lightning protection design and risk assessment. The 2022 Texas Coastal Grid Collapse, while a complex event with multiple contributing factors, served as a stark reminder of the potential consequences of inadequate lightning protection and the importance of considering site-specific conditions.

Future Research Directions

While this chapter has presented significant advancements in the field of lightning-induced flashover prediction, several avenues for future research remain:

- **More Complex Ground Models**: Expanding the modeling capabilities to include:
  - Three-dimensional variations in soil properties.
  - The presence of buried objects (pipes, cables, foundations).
  - Uneven terrain (hills, valleys).
  - More than two soil layers.
- **Frequency-Dependent Soil Properties**: Incorporating the frequency-dependent behavior of soil conductivity and permittivity into the simulations. While the current research primarily focused on constant conductivities within each layer, soil properties can vary with frequency, particularly at the higher frequencies associated with fast lightning transients.
- **Advanced Lightning Channel Models**: Exploring the use of more sophisticated lightning channel models that account for:
  - Channel tortuosity (deviations from a straight vertical path).
  - Branching of the lightning channel.
  - Non-vertical channels (building upon the work presented in Chap. 5).
- **Integration with Lightning Location Systems (LLSs)**: Developing real-time risk assessment tools that utilize data from LLSs to dynamically adjust protection system parameters or initiate proactive grid management strategies.
- **Artificial Intelligence (AI) and Machine Learning**: Applying AI and machine learning techniques to
  - Develop even more efficient and accurate flashover prediction models, potentially learning directly from large datasets of lightning parameters, ground conditions, and outage data.
  - Optimize the placement and configuration of protective devices (surge arresters, shield wires) in complex distribution networks.
  - Develop "smart grounding" systems that can adapt to changing soil conditions.

- **New Materials and Technologies**: Investigating the application of advanced materials and technologies for lightning protection, such as
  - Graphene-coated insulators with enhanced dielectric strength.
  - Superconducting fault current limiters for rapid fault interruption.
  - Advanced surge arrester technologies with improved energy absorption capabilities.
- **Validation with Field Data**: Although challenging, efforts should continue to validate these modeling techniques and analytical formulas with real-world measurements of lightning-induced voltages and flashover events on distribution lines. This would require extensive instrumentation and data collection, but it would provide the ultimate validation of the models.

By continuing to refine our understanding of lightning phenomena and their interaction with power systems, and by developing and applying advanced modeling, analysis, and mitigation techniques, we can strive toward a future with more reliable and resilient power grids, less susceptible to the disruptive and damaging effects of lightning. The research presented in this chapter, and throughout this book, represents a significant step in that direction.

References

1. E.R. Williams, J. Montanya, J. Saha, A. Guha, Lightning and climate change, in *Lightning Electromagnetics*, vol. 2. Electrical Processes and Effects (IET, 2022)
2. M. Hinow, W. Hauschild, E. Gockenbach, Lightning impulse voltage and overshoot evaluation proposed in drafts of IEC 60060-1 and future UHV testing. IEEE Trans. Dielectr. Electr. Insul. **17**(5), 1628–1634 (2010)
3. J.G. Anderson, Lightning performance of transmission lines, in *Transmission line Reference Book 345 kV and Above*, 2nd ed. (EPRI, 1982)
4. R.J.A. De Araujo, S. Kurokawa, C.M. de Seixas, B. Kordi, Lightning-induced surge in transmission towers calculated using full-wave electromagnetic analysis and the method of moments, in *13th IEEE International Conference on Industry Applications (INDUSCON)* (IEEE, 2018)
5. A. Foroughi Nematolahi, Analysis the performance of overhead line against induced voltage due to lightning in presence of different ground configuration. Post-Doctoral Research Report, Amirkabir University of Technology, Tehran, Iran (2025)
6. J.T. Whitehead et al., IEEE working group report, "Estimating lightning performance of transmission lines. II: Updates to analytical models. IEEE Trans. Power Delivery **8**(3), 1254–1267 (1993)
7. A. Borghetti, C.A. Nucci, M. Paolone, An improved procedure for the assessment of overhead line indirect lightning performance and its comparison with the IEEE Std. 1410 method. IEEE Trans. Power Delivery **22**(1), 684–692 (2007)
8. A. Borghetti, F. Napolitano, C.A. Nucci, F. Tossani, Influence of the return stroke current waveform on the lightning performance of distribution lines. IEEE Trans. Power Delivery **32**(4), 1800–1808 (2017)
9. A. Piantini, J.M. Janiszewski, The use of shield wires for reducing induced voltages from lightning electromagnetic fields. Electr. Power Syst. Res. **94**, 46–53 (2013)
10. K. Michishita, Y. Hongo, Flashover rate of 6.6-kV distribution line due to direct negative lightning return strokes. IEEE Trans. Power Delivery **27**(4), 2203–2210 (2012)

11. P. Chowdhuri, E.T.B. Gross, Voltage surges induced on overhead lines by lightning strokes. Proc. Inst. Electr. Eng. **114**(12), 1899–1907 (1967)
12. M. Darveniza, The generalized integration method for predicting impulse volt-time characteristics for non-standard wave shapes-a theoretical basis. IEEE Trans. Electr. Insul. **23**(3), 373–381 (1988)
13. W.A. Chisholm, New challenges in lightning impulse flashover modeling of air gaps and insulators. IEEE Electr. Insul. Mag. **26**(2), 14–25 (2010)
14. A. Ancajima, A. Carrus, E. Cinieri, C. Mazzetti, Optimal selection of disruptive effect models parameters for the reproduction of MV insulators volt-time characteristics under standard and non-standard lightning impulses, in *2007 IEEE Lausanne Power Tech* (IEEE, Lausanne, Switzerland, 2007), pp. 760–765
15. A. Ancajima, A. Carrus, E. Cinieri, C. Mazzetti, Behavior of MV insulators under lightning-induced overvoltages: experimental results and reproduction of volt–time characteristics by disruptive effect models. IEEE Trans. Power Delivery **25**(1), 221–230 (2010)
16. A. Foroughi Nematollahi, B. Vahidi, E.S. Ahmadi Mousavi, "Improved flashover prediction for overhead lines: considering ground stratification and insulation volt-time characteristics. IET, High Voltage, Accepted for publication (2025)
17. J. Paknahad, K. Sheshyekani, M. Hamzeh, D. Li, F. Rachidi, The influence of the slope angle of the ocean–land mixed propagation path on the lightning electromagnetic fields. IEEE Trans. Electromagn. Compat. **57**(5), 1086–1095 (2015)
18. A.F. Nematollahi, B. Vahidi, Accurate flashover assessment for coastal overhead lines based on volt-time criterion. Electr. Power Syst. Res. **239**, 111257 (2025)
19. P. Chowdhuri et al., Parameters of lightning strokes: a review. IEEE Trans. Power Delivery **20**(1), 346–358 (2005)
20. V.A. Rakov, Lightning parameters for engineering applications (keynote speech), in *2010 Asia-Pacific International Symposium on Electromagnetic Compatibility* (IEEE, Beijing, China, 2010)
21. A.R. Hileman, *Insulation Coordination for Power Systems*, 1st ed. (CRC Press, 1999)
22. CIGRE WG C4.401, Lightning parameters for engineering application. CIGRE Technical Brochure 4, Ref. 549 (2013)
23. CIGRE WG 33.01, Guide to procedures for estimating the lightning performance of transmission lines. CIGRE Technical Brochure, Ref. 063 (1991)
24. K. Sheshyekani, J. Paknahad, The effect of an ocean-land mixed propagation path on the lightning electromagnetic fields and their induced voltages on overhead lines. IEEE Trans. Power Delivery **30**(1), 229–236 (2015)
25. A. Samouëlian, I. Cousin, A. Tabbagh, A. Bruand, G. Richard, Electrical resistivity survey in soil science: a review. Soil Tillage Res. **83**(2), 173–193 (2005)
26. A.F. Nematollahi, B. Vahidi, E.S.A. Mousavi, The indirect-lightning performance of overhead distribution line considering the mixed ocean–land propagation path. IEEE Trans. Electromagn. Compat. **66**(1), 216–223 (2023)
27. M. Izadi, M.Z. A. Ab Kadir, C. Gomes, Evaluation of electromagnetic fields associated with inclined lightning channel using second order FDTD-hybrid methods. Progr. Electromagn. Res. **117**, 209–236 (2011)
28. H. Chen, Y. Du, M. Yuan, Q.H. Liu, Lightning-induced voltages on a distribution line with surge arresters using a hybrid FDTD–SPICE method. IEEE Trans. Power Delivery **33**(5), 2354–2363 (2018)
29. M. Natsui, A. Ametani, J. Mahseredjian, S. Sekioka, K. Yamamoto, FDTD analysis of distribution line voltages induced by inclined lightning channel. Electr. Power Syst. Res. **160**, 450–456 (2018)
30. H. Kohlmann, FDTD simulation of lightning electromagnetic fields: an approach with the software package MEEP. Diploma Thesis, Technische Universität Wien (2020)
31. M. Natsui, A. Ametani, J. Mahseredjian, S. Sekioka, K. Yamamoto, 3-D FDTD analysis of lightning-induced voltages in distribution lines due to inclined lightning. IEEE Trans. Electromagn. Compat. **63**(1), 189–197 (2021)

32. K. Arzag, Z.-E. Azzouz, Y. Baba, 3D-FDTD computation of lightning electromagnetic fields in the presence of a mountain and a river, in *2019 International Symposium on Lightning Protection (XV SIPDA)*, Sao Paulo, Brazil (2019)
33. T. Bdour, L. Diaz, C. Guiffaut, A. Reineix, C. Miry, P. Baraton, Emulation of an expensive FDTD code with mixed quantitative and qualitative factors for analysis of lightning-induced transient responses. IEEE Trans. Electromagn. Compat. **58**(5), 1678–1681 (2016)
34. M. Ishii, Y. Baba, Numerical electromagnetic field analysis of lightning induced voltages by moment method. IEEJ Trans. Power Energy **120**(10), 1330–1335 (2000)
35. M. Shafieipour, Z. Chen, A. Menshov, J. De Silva, V. Okhmatovski, Efficiently computing the electrical parameters of cables with arbitrary cross-sections using the method-of-moments. Electr. Power Syst. Res. **162**, 37–49 (2018)
36. "COMSOL RF Module User's Guide', COMSOL, Inc., Stockholm, Sweden, May 2016
37. F. Heidler, W. Zischank, Z. Flisowski, C. Bouquegneau, C. Mazzetti, Parameters of lightning current given in IEC 62305-background, experience and outlook, in *29th International Conference on Lightning Protection*, Uppsala, Sweden (2008)
38. V.A. Rakov et al., CIGRE technical brochure on lightning parameters for engineering applications, in *2013 International Symposium on Lightning Protection (XII SIPDA)*, Belo Horizonte, Brazil (2013)
39. A. Andreotti, F. Mottola, A. Pierno, D. Proto, Statistical features of lightning-induced voltages, in *2018 International Symposium on Power Electronics, Electrical Drives, Automation and Motion (SPEEDAM)*, Amalfi, Italy (2018)
40. A. Andreotti, R. Araneo, F. Mahmood, A. Piantini, M. Rubinstein, An analytical approach to assess the influence of shield wires in improving the lightning performance due to indirect strokes. IEEE Trans. Power Delivery **36**(3), 1491–1498 (2020)
41. A. Ackerman, P.K. Sen, C. Oertli, Designing safe and reliable grounding in AC substations with poor soil resistivity: an interpretation of IEEE Std. 80. IEEE Trans. Ind. Appl. **49**(4), 1883–1889 (2013)
42. S. Stefanescu, A. Botezan, Overview of the protection lightning standards suite EN/IEC 62305, in *2016 International Conference and Exposition on Electrical and Power Engineering (EPE)*, Iasi, Romania (2016)
43. A.F. Nematollahi, B. Vahidi, Analysis the effect of an ocean-land mixed propagation path on the induced voltage in overhead lines due to inclined lightning. Electr. Power Syst. Res. **206**, 107799 (2022)

Chapter 7
Analyzing Induced Current in the Shield of Underground Cable Due to Inclined Return Stroke Channel by 3D FEM

Abstract In this chapter the effect of three different cases is considered for the lightning channel, including the "straight inclined" channel, "bent channel", and "Tortuous channel". Straight inclined lightning is one part channel that could have different angles with *Z*-axis at *XZ* and *YZ* plane. The bent lightning is included the upper part and lower part that are joined to each other at the "bent" point. The tortuous channel is a linear piecewise approximation of realistic lightning. The impact of the channel shape, channel angle, bent height, and lightning current waveform on the induced current in the buried cable shield is investigated.

7.1 Introduction

Since years ago, research in the field of induce lightning voltage on power distribution systems and communication systems has been interesting for scientists because voltage and current induced by lightning channel can damage the equipment of these systems [1–6]. Induced voltage and current in the underground cables that are buried near the striking point can affect the wires. These induced voltages and currents can damage the cable insulation and lead to breakdown, and the power supply stops, affecting the power quality [7–9]. Accurate analysis of this electromagnetic transient can help analyze the cable failure [8, 10, 11]. In order to prevent damage to the system, accurate analysis is necessary, and grounding the cable shield for safety and EMC from a point of view is customary. As lightning strikes the ground near the underground cable, the induced current in the shield of the cable due to coupling between the shield and core can produce interference with the core; therefore, analyzing the induced current in the shield of the cable is an important issue [12–17]. Different researchers use different methods for analyzing induced current: analytical methods [14–16] or numerical methods: MoM method [17, 18], FDTD [19–21], FEM [22, 23].

In recent years, researchers have focused on the effects of soil conductivity and its structure on the induced current on underground cable. At first, researchers suggested that the ground is homogeneous and lossy [7, 11, 24]. As this assumption is not valid, new models are developed based on soil structure [12, 23, 25, 26]. In recent years,

B. Vahidi and A. F. Nematolahi, *Lightning Induced Overvoltage and Current on Distribution Power Systems*, https://doi.org/10.1007/978-981-95-3712-9_7

full-wave models based on FEM have been developed by Paknehad et al. for ground with horizontal layers, vertical layers, and frequency-dependent soil for analyzing induced current on underground cables [23, 26, 27]. Rizk et al. introduced a model based on FDTD for analyzing the induced current on the underground cable due to lightning strikes on the wind turbine tower [19].

In the last decades, researchers have focused on induced current on underground cables, but they suggested a vertical return stroke channel because a non-vertical return stroke channel leads to 3D non-homogeneous equations, which are challenging. As the return stroke channel is not vertical, the effect of an inclined return stroke channel should be investigated; therefore, this effect is studied in this chapter.

Recently, some studies on the effect of inclined return stroke channel on induced voltage on overhead lines have been conducted [1, 2, 28–30]. These studies show that the form and angle of inclination affect the induced voltage.

In this chapter, the effect of three different cases is considered for the lightning channel, including the "straight inclined" channel, "bent channel", and "Tortuous channel". Straight inclined lightning is a one-part channel that could have different angles from the *Z*-axis at the *XZ* and *YZ* planes.

The bent lightning includes the upper part and lower part that are joined to each other at the "bent" point. The tortuous channel is a linear piecewise approximation of realistic lightning. The impact of the channel shape, channel angle, bent height, and lightning current waveform on the induced current in the buried cable shield is investigated [5, 6].

The three-dimensional geometry of the case study for simulation is shown in Fig. 7.1 [5]. In this configuration, the lightning channel length, which strikes the ground, is 2 km. *D* is the buried depth of the underground cable, and *L* is the cable length. The parameters of the soil are $\sigma = 0.0017$ S/m and $\varepsilon_r = 10$. In the simulation, in order to prevent reflection from outside the boundary, the Newman natural conditions are applied to limit the computation time [31].

7.2 Lightning Current Modeling

For modeling return stroke channel current, the modified transmission line with exponential decay (MTLE) is used [19]. Equation 7.1 shows this current.

$$I(z,t) = I_0\left(0, t - \frac{z}{v}\right)e^{\frac{-z}{\lambda}} \tag{7.1}$$

where

$I(z,t)$	Return stroke channel current at time (t), at height (z)
v	Return stroke channel current velocity
$t - z/v$	Delay time of current for receiving at z height
λ	Damping coefficient
$e^{\frac{-z}{\lambda}}$	Exponential reduction of current at z height

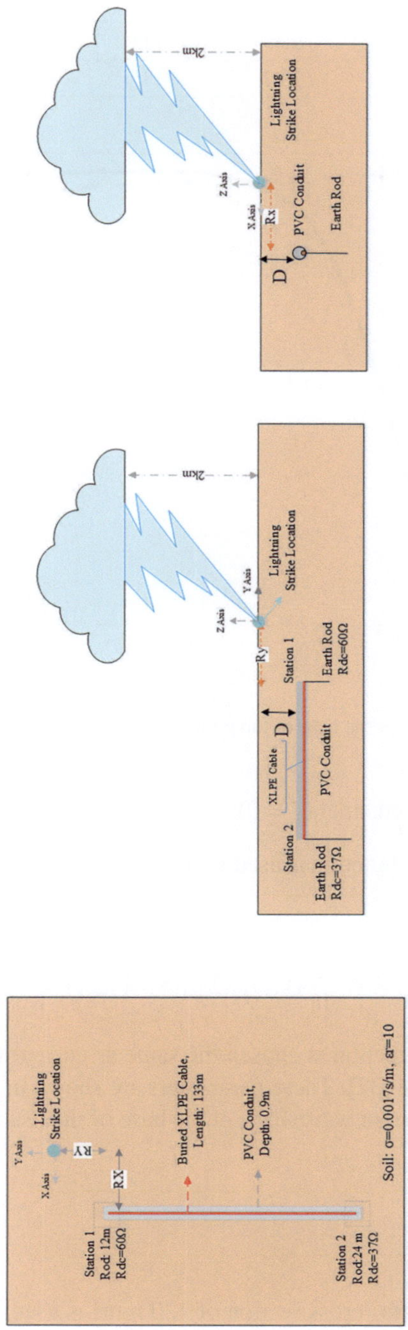

Fig. 7.1 The geometrical configuration modeling for computing the induced current in the shield of the cable

Table 7.1 The parameters of the current channel base according to IEC 62,305

Current wave form	I_0(kA)	$\tau_1(\mu s)$	$\tau_2(\mu s)$	n_1	η_1
0.25/100	50	0.454	143	10	0.993
1/200	50	1.82	285	10	0.986

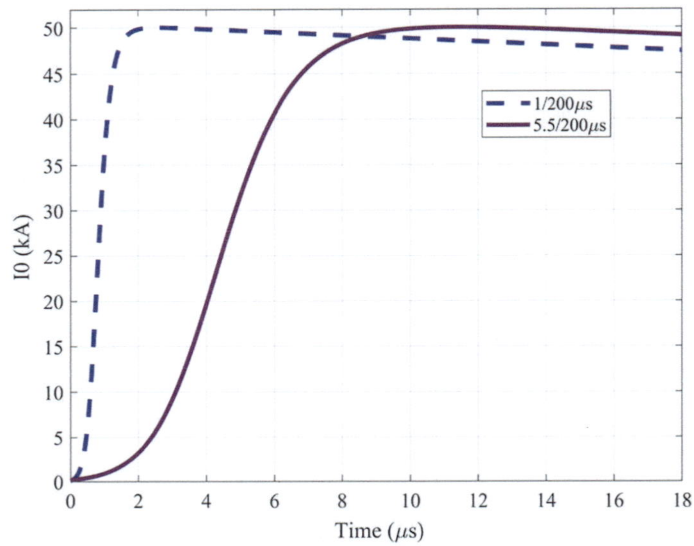

Fig. 7.2 The current waveshape at the base of the return stroke channel

$I_0(0, t)$ Current at base of channel ($z = 0$).

In this chapter, Hiedler's function is used for the current at the base of the channel (Eq. 7.2).

$$I(0,t) = \frac{I_0}{\eta_1}\frac{(t/\tau_1)^{n_1}}{1+(t/\tau_1)^{n_1}}\exp\left(\frac{-t}{\tau_2}\right) \tag{7.2}$$

Parameters of Heidler function is chosen the same as current waveshape of IEC 62,305 for 50 kA amplitude [32]. These parameters are shown in Table 7.1.

Figure 7.2 shows the current waveshape at the base of the return stroke channel.

7.3 Cable Modeling

A 30 kV cable with conductor cross section of 120 mm^2 is used. The cable configuration is shown in Fig. 7.3. The cable contains four main parts; conductor, XLPE insulation, cable shield and PVC sheath.

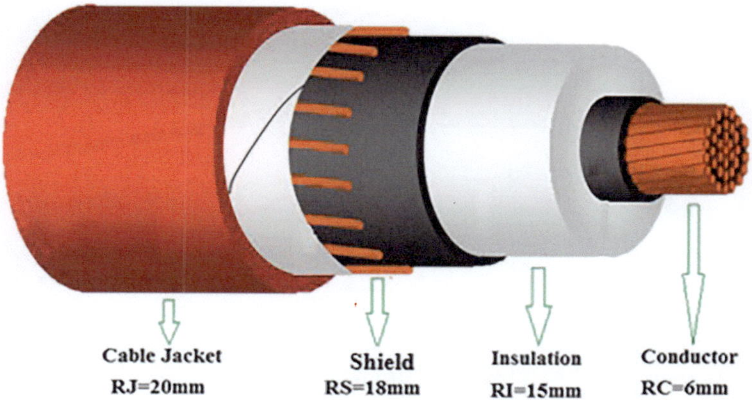

Fig. 7.3 The simulated cable

The cable is buried at 90 cm depth; the length of cable is 133 m and cable grounded at both ends (station 1 and station 2). The grounding resistance at station 1 is 37 Ohms and 60 Ohms at station 2. As FEM introduces a full wave solution for Maxwell equations, therefore coupling between different parts of cable and return stroke channel is considered.

7.4 Simulation and Results

In this section series of simulation conducted for evaluating the effect of geometrical configuration of return stroke channel on induced current on shield of cable. Three different cases are considered for the lightning channel, including the "straight inclined" channel, "bent channel", and "Tortuous channel". Two current waveshape as Table 7.1 with current amplitude of 50 kA is considered for the simulation. Cable length is 133 m, $Rx = 94$ m and $Ry = 230$ m (Fig. 7.1) and $\sigma = 0.0017$ S/m and $\varepsilon_r = 10$.

7.4.1 Comparison of Results with Measurement Data

For confirmation of simulations results of vertical return stroke channel are compared with results of transmission line model and measured values [33, 34]. The geometry of simulation is shown in Fig. 7.1. Measured values are from ICLRT center in Florida (details for measurement are shown in reference [34]). As can see in Fig. 7.4 the risetime and amplitude of evaluated current and measured current are closed to each other, but the fall times are different.

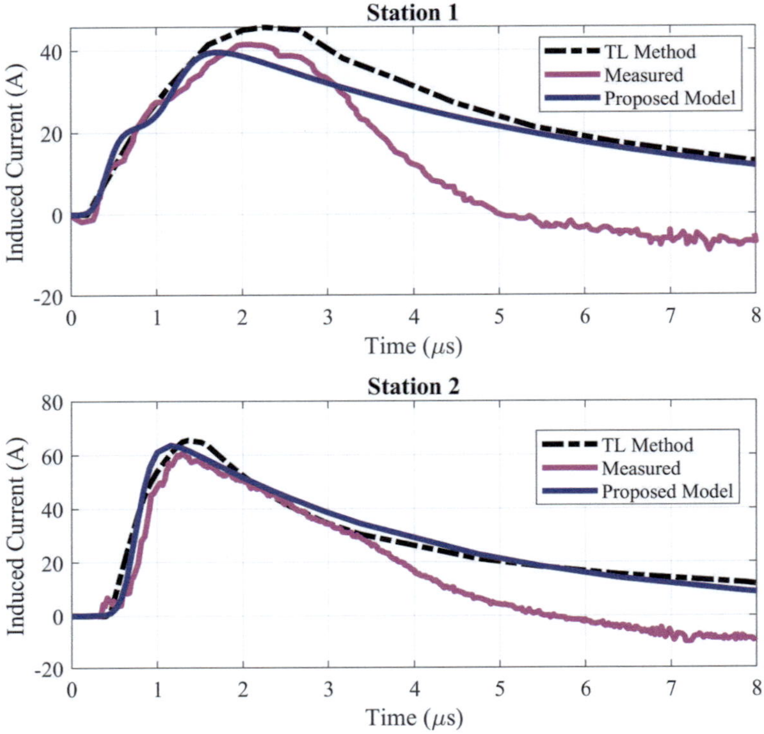

Fig. 7.4 A comparison of the evaluated induced current and measured induced current at stations 1 and 2 (Fig. 7.1) shows that the parameters and details can be found in [34]

7.4.2 Straight Inclined Return Stroke Channel

Figure 7.5 shows the geometry of the simulation for the inclined return stroke channel. The angle between the inclined channel and the *Z* axis in the *XZ* plane is φ, and the angle between the inclined channel and the *Z* axis in the *YZ* plane is θ.

By variation of φ and θ between $-\ 45°$ and $+\ 45°$ in seven steps, the induced currents on both stations of the cable are evaluated.

Simulation conducted for two lightning waveshapes, 1/200 µs and 5.5/200 µs, and induced currents at stations 1 and 2 for these lightning currents are shown in Figs. 7.6 and 7.7.

The angle θ has much more influence on the amplitude of the induced voltage in comparison to φ. For $\theta > 0°$, the induced current, in contrast to the vertical return stroke channel, increases because the distance between the inclined return stroke channel and the cable stations decreases. The amplitude of the induced current at station 1 for $\theta = 45°$ is about 200% of the induced current due to the vertical return stroke channel. For $\theta < 0°$, the induced current in comparison to the vertical return stroke channel decreases because the distance between the inclined return stroke

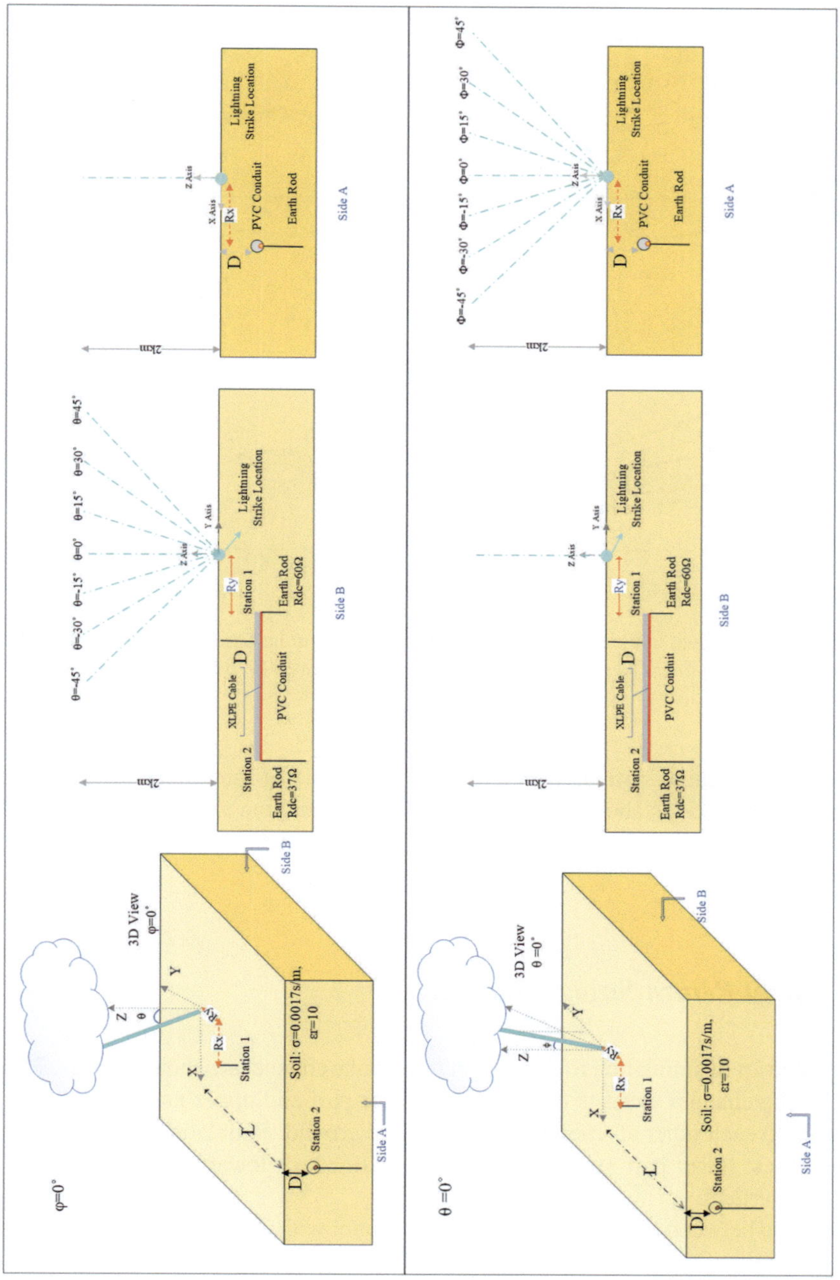

Fig. 7.5 The geometry of simulation for evaluating the induced current in the shield of an underground cable for a straight inclined return stroke channel

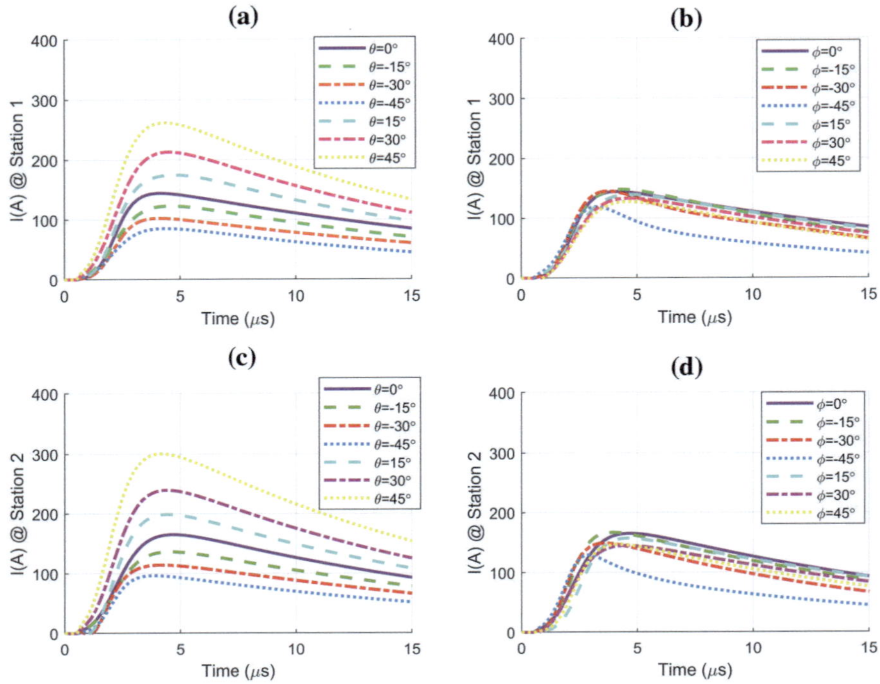

Fig. 7.6 Induced current in the shield of underground cable due to inclined return stroke channel (lightning waveshape: 1/200 μs) at stations 1 and 2 for different cases (**a**: $\varphi = 0°$, **b**: $\theta = 0°$, **c**: $\varphi = 0°$, **d**: $\theta = 0°$)

channel and the cable stations increases. The amplitude of the induced current at station 1 for $\theta = 45°$ is about 50% of the induced current due to the vertical return stroke channel. The angle φ has less influence on the amplitude of the induced voltage in comparison to θ.

7.4.3 Bent Return Stroke Channel

The geometries of simulation for the bent return stroke channel are shown in Figs. 7.8 and 7.9. The channel consists of two parts (lower part and upper part), which are connected to each other at a height of "Ha" above ground. Simulation is conducted for two cases: lower part vertical, upper part inclined, or lower part inclined and upper part vertical.

The angle between the inclined part of the channel with Z axis in the XZ plane is φ and the angle of the inclined part of channel with Z axis in the YZ plane is θ.

By variation of φ and θ between $-45°$ and $+45°$ in seven steps, and for Ha = 50 m, 100 m, 250 m, 500 m the induced currents in the shield of cable are evaluated.

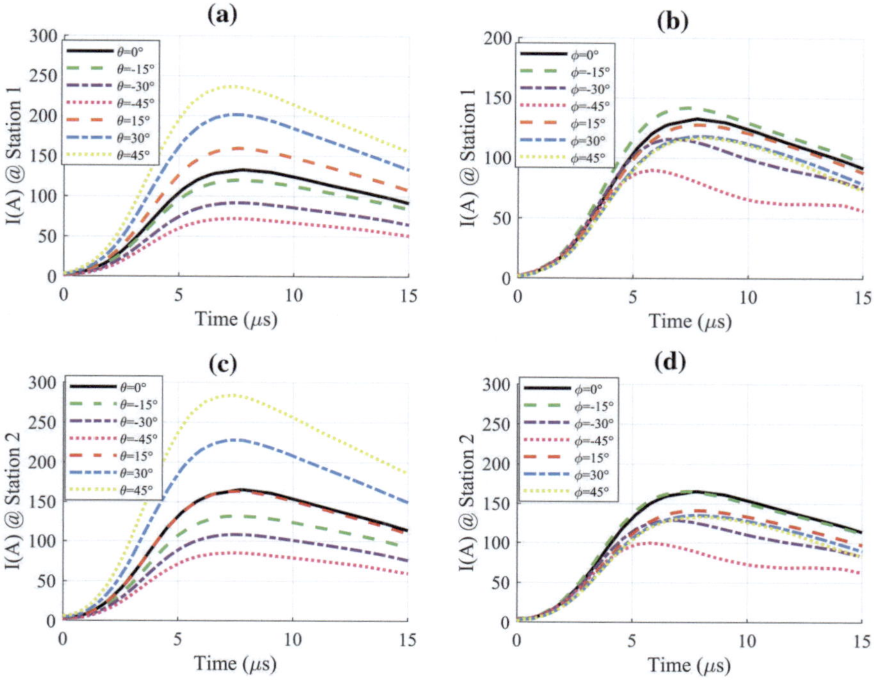

Fig. 7.7 Induced current in the shield of underground cable due to inclined return stroke channel (lightning waveshape: 5.5/200 μs) at stations 1 and 2 for different cases (**a**: $\varphi = 0°$, **b**: $\theta = 0°$, **c**: $\varphi = 0°$, **d**: $\theta = 0°$)

Figures 7.10 and 7.11 show the induced currents in the shield of cable at station 1 for different values of Ha, φ and θ for 1/200 μs current waveshape and the geometry is as Fig. 7.8. The peak of the induced current is shown in Fig. 7.12.

By comparison of results for different Ha values the effect of θ and φ can be resulted.

From these results we can judge that θ has not much influence on the rise time and fall time of the induced current. In Fig. 7.11 the influence of φ angle on the induced current are shown. For $\varphi < 0°$ when Ha is 50 or 100 m the fall time is reduced.

Simulation results for geometry of Fig. 7.9 are shown in Figs. 7.13 and 7.14. From Fig. 7.13 can judge that the θ angle and Ha do not have much influence on the risetime and fall time of the induced current waveshape.

From Fig. 7.14, we can judge that the φ angle has less influence on the induced current amplitude compared to the θ angle.

Also, it can be judged that by increasing Ha, the influence of the bent angle on the induced current decreases, for Ha > 250 m, the induced current approximately is the same as the induced current due to the inclined channel (Ha = 2000 m).

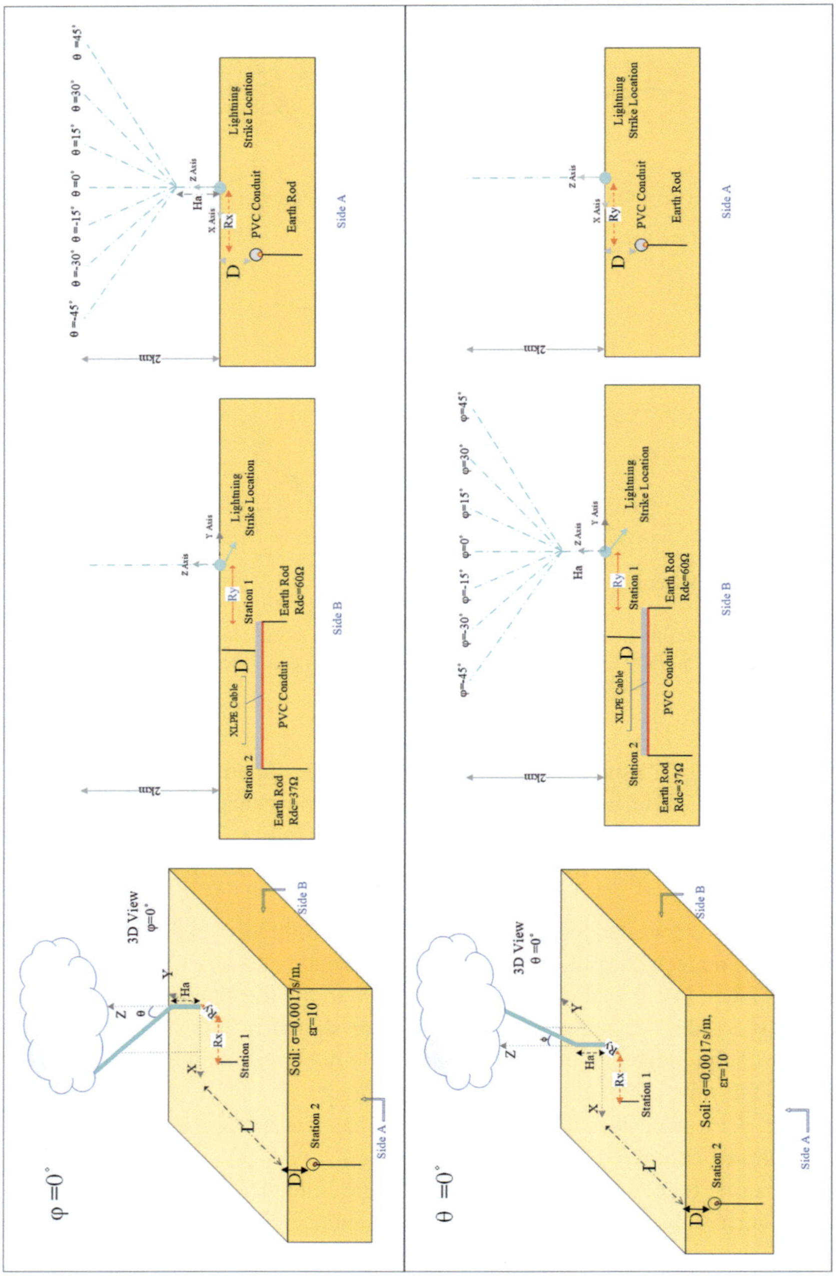

Fig. 7.8 The geometry of the simulation for evaluating induced current in the shield of underground cable for bent return stroke channel (the The lower part is vertical, and the upper part is inclined

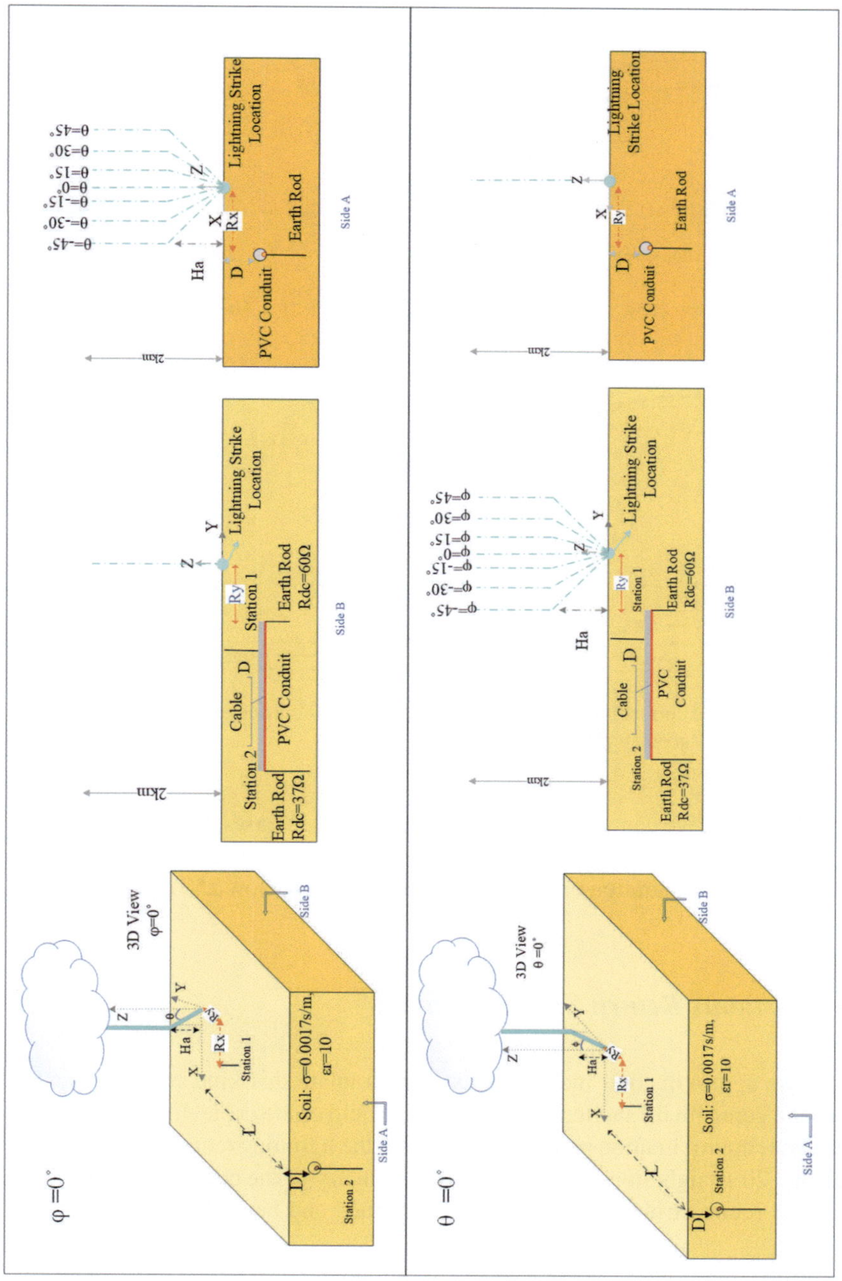

Fig. 7.9 The geometry of the simulation for evaluating the induced current in the shield of underground cable for bent return stroke channel (the The lower part is inclined, and the upper part is vertical

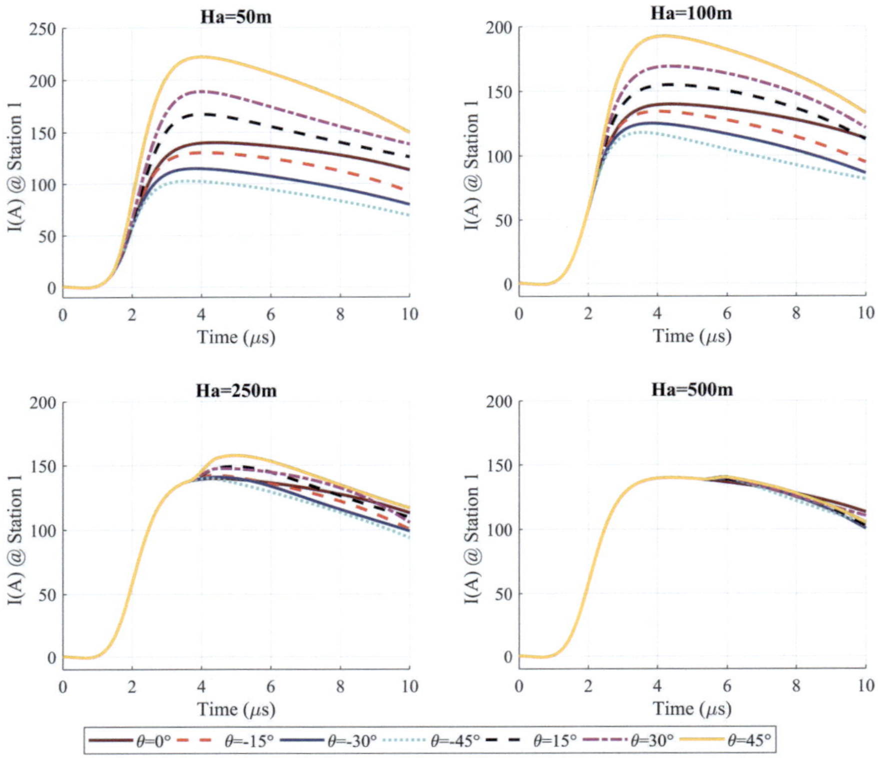

Fig. 7.10 The induced current in the shield of cable at station 1 for different Ha values and $\varphi = 0°$ (waveshape: 1/200 μs, geometry: Fig. 7.8)

From Fig. 7.14, we can judge that for $\varphi > 0°$, if Ha increases, the risetime of the current decreases. The simulation results show that the induced current is related to the geometrical configuration of the return stroke channel below 250 m over ground.

7.4.4 *Tortuous Return Stroke Channel*

A piecewise approximation method is used to simulate this kind of return stroke channel and compute the induced current on the shield of the cable. In order to apply the piecewise approximation method, the photo of the lightning return stroke channel is used, and 20 straight lines approximate the geometry of the channel. In Fig. 7.15, the tortuous return stroke channel, its equal inclined and bent channels are shown [5].

The results for tortuous, inclined, bent, and straight vertical return stroke channels are shown in Fig. 7.16 [5]. In these simulations:

For inclined return stroke channel: $\theta = 17°$, $\varphi = 15°$.

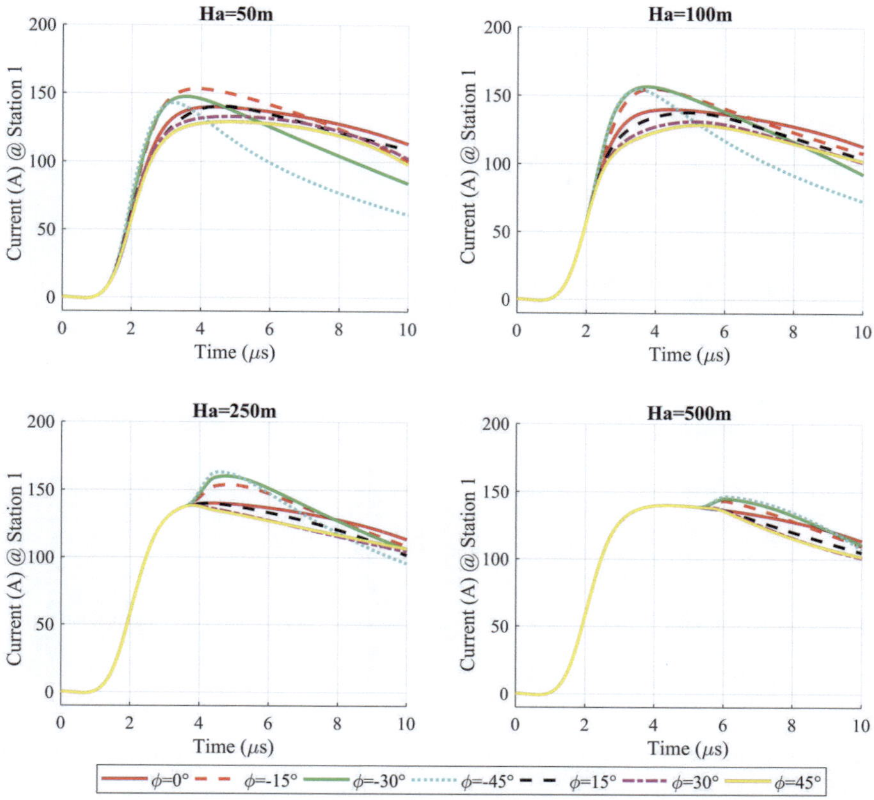

Fig. 7.11 The induced current in the shield of cable at station 1 for different Ha values and $\theta = 0°$ (waveshape: 1/200 μs, geometry: Fig. 7.8)

For bent return stroke channel: Ha = 350 m, lower part: $\theta = 26°$, $\varphi = 34°$, upper part: $\theta = -32°$, $\varphi = 10°$.

Lightning current waveshape: 1/200 μs, 50 kA and the geometrical configuration of the simulation is as shown in Fig. 7.1.

From Fig. 7.16, it can be seen that there is some oscillation on the waveshape of the induced current due to the piecewise approximation. From Fig. 7.16, we can judge that the induced current of the bent channel is closer to the piecewise approximation. The difference between the peak of induced currents due to the piecewise approximation and the bent channel is 3%. The difference between the peak of induced currents due to the piecewise approximation and the inclined channel is 11.5%. The difference between the peak of induced currents due to the piecewise approximation and the straight vertical channel is 21%.

However, the results of induced currents on the shield of cables due to an inclined return stroke channel show a good evaluation of this phenomenon. The bent return stroke channel shows a better approximation for natural lightning, because the lower

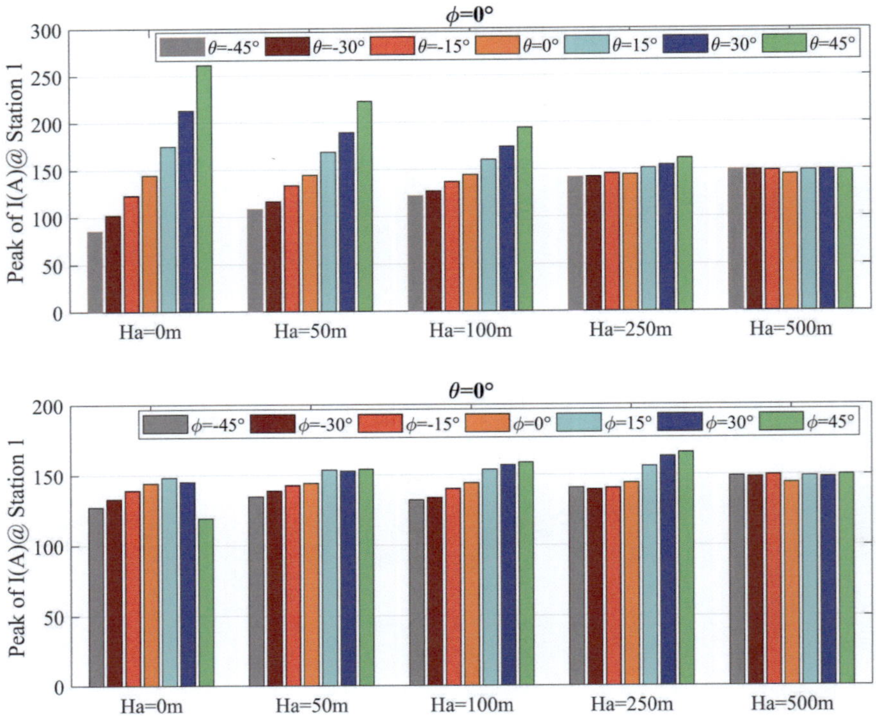

Fig. 7.12 The peak of induced current in the shield of cable at station 1 for different Ha values for the geometry of Fig. 7.8

part of the bent channel shows a better approximation for the lower part of natural lightning. As the lower part of the return stroke channel has much influence on the induced current, the results of the bent channel are closer to the actual measured values.

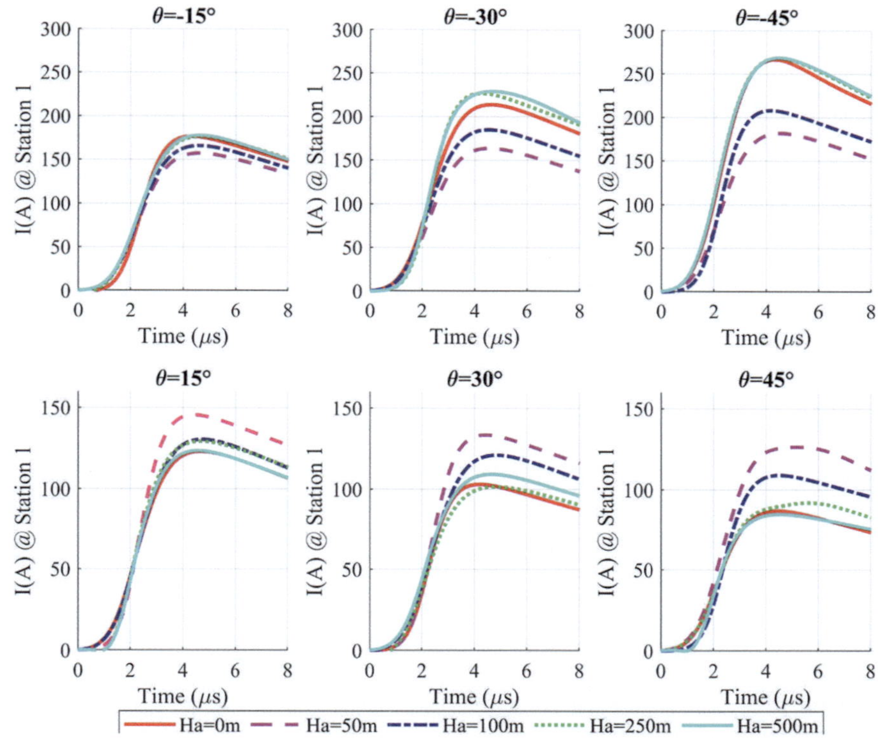

Fig. 7.13 The induced current in the shield of cable at station 1 for different Ha values and $\varphi = 0°$ (waveshape: 1/200 μs, geometry: Fig. 7.9)

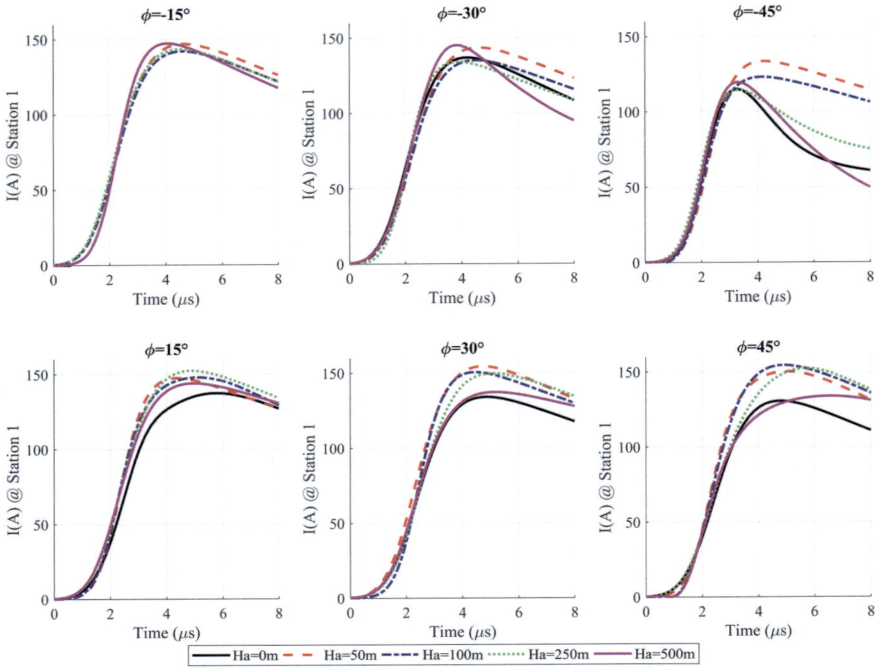

Fig. 7.14 The induced current in the shield of cable at station 1 for different Ha values and $\theta = 0°$ (waveshape: 1/200 μs, geometry: Fig. 7.9)

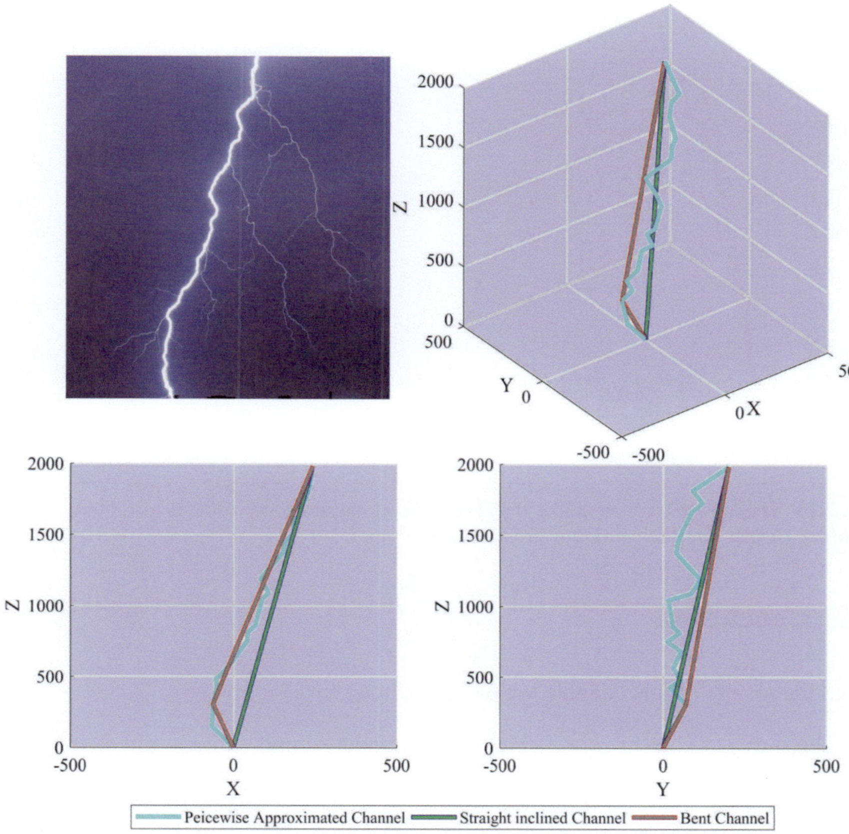

Fig. 7.15 The photo of the return stroke channel and its equal inclined and bent return stroke channels

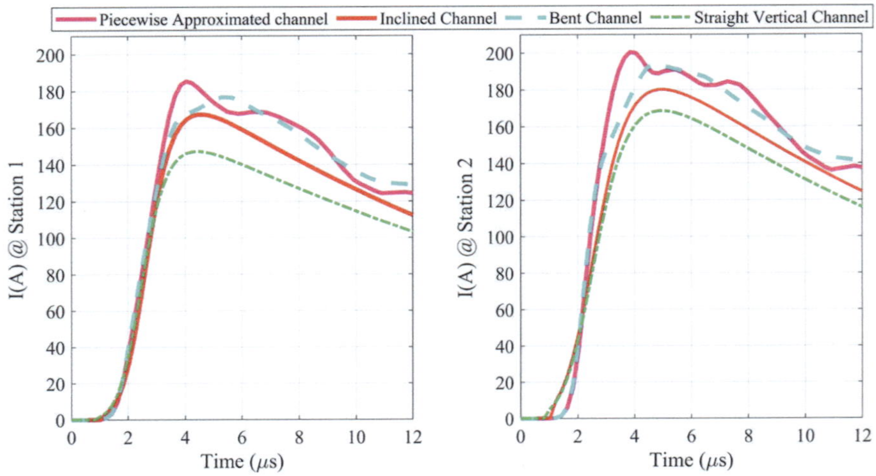

Fig. 7.16 The induced current on the shield of cable at stations 1 and 2 due to a 1/200 μs, 50 kA lightning current for different return stroke channel shapes

7.5 Conclusion

From the results of this chapter, we can conclude:

1. The induced current in the shield of the underground cable is affected by the lightning current waveshape, the return stroke channel configuration, its direction, and the angle of the return stroke channel. The amplitude of the induced current in the shield of the underground cable for a non-vertical return stroke channel can be 50–200% of the induced current due to a vertical channel.
2. For a 1/200 μs lightning current waveshape, the induced current in the shield of the underground cable is affected by the geometrical configuration of the return stroke channel at a height below 250 m.
3. The tortuous return stroke channel can be modeled by two parts, where the lower part (under 250 m height) should have a good match with the natural lightning return stroke channel.
4. The results of analysis show that if the lightning current has a shorter risetime, the induced current in the shield of the underground cable is more affected by the return stroke channel configuration.

References

1. A.F. Nematollahi, B. Vahidi, The effect of the inclined lightning channel on electromagnetic fields and the induced voltages on overhead lines. Electr. Eng. **103**, 3163–3176 (2021)

2. A.F. Nematollahi, B. Vahidi, Analysis the effect of an ocean-land mixed propagation path on the induced voltage in overhead lines due to inclined lightning. Electric Power Systems Research. **206**, 107799 (2022)
3. W. Hou, M. Azadifar, M. Rubinstein, F. Rachidi, Q. Zhang, On the propagation of lightning-radiated electromagnetic fields across a mountain. IEEE Trans. Electromagn. Compat. **62**, 2137–2147 (2020)
4. M. Nazari, A. Aghabarati, R. Moini, S. Fortin, Y. Li, F.P. Dawalibi, Evaluating hazards of lightning strike on stationary vehicles under overhead lines, in *2017 IEEE 5th International Symposium on Electromagnetic Compatibility (EMC-Beijing)* (IEEE, 2017), pp. 1–6
5. A.F. Nematollahi, Investigation of induced overvoltage caused by lightning in distribution networks considering environmental and soil conditions, Ph.D. thesis, Amirkabir University of Technology, Tehran, Iran, 2023
6. A.F. Nematollahi, Analysis of performance of overhead lines against lightning induced voltage in presence of different ground structres, Post Doctoral Report, Amirkabir University of Technology, Tehran, Iran, 2025
7. L.-M. Zhou, S.A. Boggs, Effect of shielded distribution cables on lightning-induced overvoltages in a distribution system. IEEE Trans. Power Delivery **17**, 569–574 (2002)
8. Q. Yang, J. Tao, Y. He, B. Zhang, Analysis and suppression measures of lightning transient overvoltage in the signal cable of wind turbines. Wind Energy **21**, 211–225 (2018)
9. J.D. Nordgard, C. Chen, Lightning-induced transients on buried shielded transmission lines. IEEE Trans. Electromagn. Compat. 171–181 (1797)
10. A. De Conti, O.E.S. Leal, Lightning-induced voltage calculations on an overhead insulated cable, in *2019 International Symposium on Lightning Protection (XV SIPDA)* (IEEE, 2019), pp. 1–4
11. A. Borghetti, M. Marzinotto, C. Mazzetti, C.A. Nucci, M. Paolone, Insulation coordination of MV cables against lightning-induced overvoltages generated by LEMP-coupled overhead lines, in *28th International Conference on Lightning Protection (ICLP)* (2006)
12. H. Tanaka, Y. Baba, C.F. Barbosa, Effect of shield wires on the lightning-induced currents on buried cables. IEEE Trans. Electromagn. Compat. **58**, 738–746 (2016)
13. A. Ametani, N. Taki, D. Miyazaki, N. Nagaoka, S. Okabe, Lightning surges on a control cable incoming through a grounding lead. IEEJ Trans. Power Energy **127**, 267–275 (2007)
14. M. Zhu, S. Liu, H. Zhou, P. Sun, Y. Xie, J. Tian, Experimental study of EMP coupling to long shielded cable. High Power Laser Part. Beams **13**, 761–765 (2001)
15. E. Petrache, M. Paolone, F. Rachidi, C.A. Nucci, V. Rakov, M. Uman, D. Jordan, K. Rambo, J. Jerauld, M. Nyffeler, Lightning-induced currents in buried coaxial cables: a frequency-domain approach and its validation using rocket-triggered lightning. J. Electrostat. **65**, 322–328 (2007)
16. F. Rachidi, A review of field-to-transmission line coupling models with special emphasis on lightning-induced voltages on overhead lines. IEEE Trans. Electromagn. Compat. **54**, 898–911 (2012)
17. B. Zhang, X. Cui, Z. Zhao, J. He, L. Li, Numerical analysis of the influence between large grounding grids and two-end grounded cables by the moment method coupled with circuit equations. IEEE Trans. Power Delivery **20**, 731–737 (2005)
18. M. Shafieipour, Z. Chen, A. Menshov, J. De Silva, V. Okhmatovski, Efficiently computing the electrical parameters of cables with arbitrary cross-sections using the method-of-moments. Electr. Power Syst. Res. **162**, 37–49 (2018)
19. M.E.M. Rizk, F. Mahmood, M. Lehtonen, E.A. Badran, M.H. Abdel-Rahman, Investigation of lightning electromagnetic fields on underground cables in wind farms. IEEE Trans. Electromagn. Compat. **58**, 143–152 (2016)
20. Y. Baba, N. Tanabe, N. Nagaoka, A. Ametam, Transient analysis of a cable with low-conducting layers by a finite-difference time-domain method. IEEE Trans. Electromagn. Compat. **46**, 488–493 (2004)
21. A. Tatematsu, A technique for representing coaxial cables for FDTD-based surge simulations. IEEE Trans. Electromagn. Compat. **57**, 488–495 (2015)

22. J. Paknahad, K. Sheshyekani, F. Rachidi, M. Paolone, A. Mimouni, Evaluation of lightning-induced currents on cables buried in a lossy dispersive ground. IEEE Trans. Electromagn. Compat. **56**, 1522–1529 (2014)
23. J. Paknahad, K. Sheshyekani, F. Rachidi, M. Paolone, Lightning electromagnetic fields and their induced currents on buried cables. Part II: the effect of a horizontally stratified ground. IEEE Trans. Electromagn. Compat. **56**, 146–1154 (2014)
24. C.A. Charalambous, N. Kokkinos, N. Christofides, M.Z.A. Ab Kadir, C. Gomes, A simulation tool to assess the lightning induced over-voltages on dc cables of photovoltaic installations, in *2014 International Conference on Lightning Protection (ICLP)* (IEEE, 2014), pp. 1571–1576
25. K. Sheshyekani, J. Paknahad, The effect of an ocean-land mixed propagation path on the lightning electromagnetic fields and their induced voltages on overhead lines. IEEE Trans. Power Delivery **30**, 229–236 (2015)
26. J.O. Paulino, C.F. Barbosa, W. do Couto Boaventura, Lightning-induced current in a cable buried in the first layer of a two-layer ground. IEEE Trans. Electromagn. Compat. **56**, 956–963 (2013)
27. J. Paknahad, K. Sheshyekani, F. Rachidi, Lightning electromagnetic fields and their induced currents on buried cables. Part I: the effect of an ocean–land mixed propagation path. IEEE Trans. Electromagn. Compat. **56**, 1137–1145 (2014)
28. A. Andreotti, A. Piantini, A. Pierno, R. Rizzo, Lightning-induced voltages on complex power systems by using CiLIV: the effects of channel tortuosity. IEEE Trans. Power Delivery **33**, 680–688 (2017)
29. A. Andreotti, V.A. Rakov, L. Verolino, Exact and approximate analytical solutions for lightning-induced voltage calculations. IEEE Trans. Electromagn. Compat. **60**, 1850–2185 (2018)
30. M. Natsui, A. Ametani, J. Mahseredjian, S. Sekioka, K. Yamamoto, FDTD analysis of distribution line voltages induced by inclined lightning channel. Electr. Power Syst. Res. **160**, 450–456 (2018)
31. COMSOL, COMSOL RF Module User's Guide, COMSOL, Inc., Stockholm, Sweden, 2016
32. IEC Standard 62305, Protection against lightning Part 1: General Principles, IEC Switzerlan, 2011
33. E. Petrache, F. Rachidi, M. Paolone, C.A. Nucci, V.A. Rakov, M.A. Uman, Lightning induced disturbances in buried cables-part I: theory. IEEE Trans. Electromagn. Compat. **47**, 498–508 (2005)
34. M. Paolone, E. Petrache, F. Rachidi, C.A. Nucci, V.A. Rakov, M.A. Uman, D. Jordan, K. Rambo, J. Jerauld, M. Nyffeler, Lightning induced disturbances in buried cables-part II: experiment and model validation. IEEE Trans. Electromagn. Compat. **47**, 509–520 (2005)

Index

B. Vahidi and A. F. Nematolahi, *Lightning Induced Overvoltage and Current on Distribution Power Systems*, https://doi.org/10.1007/978-981-95-3712-9

S

T

U

V

W

MIX
Papier aus verantwortungsvollen Quellen
Paper from responsible sources
FSC® C105338

If you have any concerns about our products,
you can contact us on
ProductSafety@springernature.com

In case Publisher is established outside the EU,
the EU authorized representative is:
Springer Nature Customer Service Center GmbH
Europaplatz 3, 69115 Heidelberg, Germany

Printed by Libri Plureos GmbH
in Hamburg, Germany